Ökonometrie und Unternehmensforschung

Econometrics and Operations Research

XXI

Herausgegeben von *Edited by*

M. Beckmann, München/Providence R. Henn, Karlsruhe
A. Jaeger, Bochum W. Krelle, Bonn H. P. Künzi, Zürich
K. Wenke, Zürich Ph. Wolfe, New York

Geschäftsführende Herausgeber *Managing Editors*

W. Krelle H. P. Künzi

Peter Kall

Stochastic Linear Programming

Springer-Verlag Berlin Heidelberg New York 1976

Peter Kall
Institute for Operations Research
and Mathematical Methods in Economics, University of Zurich

AMS Subject Classifications (1970): 28A20, 60E05, 90–02, 90C05, 90C15, 90C20, 90C25

ISBN-13: 978-3-642-66254-6 e-ISBN-13: 978-3-642-66252-2
DOI: 10.1007/978-3-642-66252-2

Library of Congress Cataloging in Publication Data
Kall, Peter. Stochastic linear programming. (Ökonometrie und Unternehmesforschung; 21). Bibliography: p. Includes index. 1. Linear programming. 2. Stochastic processes. I. Title. II. Series.
HB143.K35 519.7'2. 75-30602.

Softcover reprint of the hardcover 1st edition 1976

Preface

Today many economists, engineers and mathematicians are familiar with linear programming and are able to apply it. This is owing to the following facts: during the last 25 years efficient methods have been developed; at the same time sufficient computer capacity became available; finally, in many different fields, linear programs have turned out to be appropriate models for solving practical problems.

However, to apply the theory and the methods of linear programming, it is required that the data determining a linear program be fixed known numbers. This condition is not fulfilled in many practical situations, e. g. when the data are demands, technological coefficients, available capacities, cost rates and so on. It may happen that such data are random variables. In this case, it seems to be common practice to replace these random variables by their mean values and solve the resulting linear program. By 1960 various authors had already recognized that this approach is unsound: between 1955 and 1960 there were such papers as "Linear Programming under Uncertainty", "Stochastic Linear Programming with Applications to Agricultural Economics", "Chance Constrained Programming", "Inequalities for Stochastic Linear Programming Problems" and "An Approach to Linear Programming under Uncertainty".

The aim of this book is to give some insight into this challenging field which has to be understood as a special subject of planning under uncertainty. A complete collection of results obtained so far did not seem entirely appropriate, and my preference led me to choose those topics and results which can be handled more or less systematically within a certain theoretical framework. This does not imply a value judgement on topics and results which are not reported. In the bibliography I have cited only those papers which were really used in the writing of the text. A fairly comprehensive bibliography on stochastic linear programming can be obtained by taking the union of the bibliographies of the books cited and all references found by starting with the papers on stochastic programming listed at the end.

It is assumed that the reader is familiar with elementary real analysis and linear algebra. It would be helpful if he were also acquainted with optimization theory as well as basic measure theory and probability theory. With regard to the latter requirements, and also to avoid terminological confusions, I have included a collection of the most important definitions and results (Chapter 0.), to which I refer later on. Beyond these prerequisites, every assertion is proved which, in my opinion, leads to a better understanding of the results, the difficulties and the unsolved problems.

I am indebted to dipl. math. B. Finkbeiner, Dr. K. Hässig, Dr. M. Köhler, Dr. K. Marti, Dr. R. J. Riepl and especially to Prof. Dr. W. Vogel and Prof. Dr. R. Wets for their helpful comments and suggestions. Nevertheless I am responsible for every mistake left, and I shall appreciate every constructive criticism. I also owe thanks to Mrs. E. Roth for typing the manuscript and giving linguistic support. Finally I have to acknowledge the extraordinary patience of the editors and Springer-Verlag.

Contents

Chapter 0. Prerequisites

1. Linear Programming

Linear Programs are special mathematical programming problems. We understand by a *mathematical programming problem* in the Euclidean space $\mathbb{R}^n$ the optimization of a given real-valued function — *the objective function* — on a given subset of $\mathbb{R}^n$, the so-called *feasible set*. A mathematical programming problem is called a *linear program* if its objective function is a linear functional on $\mathbb{R}^n$ and if the feasible set can be described as the intersection of finitely many halfspaces and at most finitely many hyperplanes in $\mathbb{R}^n$. Hence the feasible set of a linear program may be represented as the solution set of a system of finitely many linear inequalities and, at most, finitely many linear equalities — the so-called (linear) *constraints* — in a finite number of variables.

For $f \in \mathbb{R}^n$, $\alpha \in \mathbb{R}$ it is evident that

$$\{x \mid x \in \mathbb{R}^n, f'x \leq \alpha\} = \{x \mid x \in \mathbb{R}^n, \tau \in \mathbb{R}, f'x + \tau = \alpha, \tau \geq 0\},$$

where f' is the transpose of f and $f'x$ is the usual scalar product of f and x. From this relation, and from the fact that every real number may be represented as the difference of two nonnegative real numbers, it follows that every linear program may be written in the standard formulation

$$\min\{c'x \mid Ax = b, x \geq 0\}, \tag{1}$$

where $c \in \mathbb{R}^n$, $b \in \mathbb{R}^m$ are constant and A is a constant real $(m \times n)$-matrix, $x \in \mathbb{R}^n$ is the vector of *decision variables* and $x \geq 0$ stands for $x_i \geq 0$, $i = 1, \ldots, n$. We understand by a *solution* of a linear program a feasible $\hat{x}$ such that $c'\hat{x} \leq c'x$ for all $x \in \{x \mid Ax = b; x \geq 0\}$.

The question, under which conditions the feasible set of a linear program is nonempty, is answered by

Theorem 1 (Farkas' Lemma). $\{x \mid Ax = b, x \geq 0\} \neq \emptyset$ *if and only if* $\{u \mid A'u \geq 0\} \subset \{u \mid b'u \geq 0\}$.

Here the prime at A and b again indicates transposition.

An immediate consequence of Farkas' Lemma is

Theorem 2. *If there is a real constant γ such that $c'x \geq \gamma$ for all $x \in \{x \mid Ax = b, x \geq 0\} \neq \emptyset$, then the linear program* $\min\{c'x \mid Ax = b, x \geq 0\}$ *has a solution.*

An important concept in linear programming is that of feasible basic solutions. We call $\tilde{x}$ a *feasible basic solution* of (1) if $\tilde{x} \in \{x \mid Ax = b, x \geq 0\}$ and if the columns of A corresponding to the nonzero components of $\tilde{x}$ are linearly independent. Obviously the set $\mathfrak{A} = \{x \mid x$ is a feasible basic solution of (1)$\}$ is finite.

If we define the *convex polyhedron* generated by $\mathfrak{A}$ as

$$\mathfrak{P}=\{x\,|\,x=\sum_{i=1}^{r}\lambda_i z^i, z^i\in\mathfrak{A}, \lambda_i\geq 0, \sum_{i=1}^{r}\lambda_i=1\}$$

and observe that $\mathfrak{K}=\{y\,|\,Ay=0, y\geq 0\}$ is a *convex polyhedral cone*, i.e. if $y^1\in\mathfrak{K}$ and $y^2\in\mathfrak{K}$, then $\lambda_1 y^1+\lambda_2 y^2\in\mathfrak{K}$ for all $\lambda_1\geq 0, \lambda_2\geq 0$ and $\mathfrak{K}$ is generated by a finite set, then we may state

Theorem 3. $\{x\,|\,Ax=b, x\geq 0\}=\mathfrak{P}+\mathfrak{K}=\{x\,|\,x=z+y, z\in\mathfrak{P}, y\in\mathfrak{K}\}$.

In other words: The feasible set of a linear program is the direct (or algebraic) sum of a convex polyhedron and a convex polyhedral cone. Sets of this type are usually called *convex polyhedral sets.* From the representation of feasible solutions given by Th. 3, follows immediately

Theorem 4. *The linear program* (1) *has a solution if and only if* $\{x\,|\,Ax=b, x\geq 0\}\neq\emptyset$ *and* $c'y\geq 0$ *for all* $y\in\{y\,|\,Ay=0, y\geq 0\}$.

Furthermore, we may conclude

Theorem 5. *If the linear program* (1) *has a solution, then at least one of the feasible basic solutions is also a solution.*

Therefore, if we want to determine a solution of a linear program, we may restrict ourselves to the investigation of a finite number of points, namely the feasible basic solutions. This is done in the well-known *Simplex method.* To describe the essential parts of this method, we assume without loss of generality that in (1) the matrix A has rank $m\leq n$. Then A contains subsets of columns $\{A_{i_1},\ldots,A_{i_m}\}$, which are linearly independent and hence constitute bases of $\mathbb{R}^m$. Such a basis, written as a matrix $B=(A_{i_1},\ldots,A_{i_m})$, is called a *feasible basis* if $B^{-1}b\geq 0$.
If D is the matrix of the $n-m$ columns, which are not contained in B, and if $\tilde{x}=\{x_{i_1},\ldots,x_{i_m})$ and y is the $(n-m)$-tuple of x-variables corresponding to the columns of D, then $Ax=b$ is equivalent to $B\tilde{x}+Dy=b$. Hence, the vector of *basic variables* $\tilde{x}$ depends on the vector of *nonbasic variables* y as follows:

(2) $$\tilde{x}=B^{-1}b-B^{-1}Dy.$$

If B is a feasible basis and if we choose $y=0$, then we have a feasible basic solution, where the basic variables have the values of the components of $B^{-1}b$. If we reorder the components of c into an m-tuple $\tilde{c}$ and an $(n-m)$-tuple d corresponding to the reordering of the components of x in $\tilde{x}$ and y, then, from (2), we get for the objective function

(3) $$\begin{aligned} c'x&=\tilde{c}'\tilde{x}+d'y\\ &=\tilde{c}'B^{-1}b+(d'-\tilde{c}'B^{-1}D)y. \end{aligned}$$

Starting with a feasible basis B and the corresponding feasible basic solution given by $y=0$, $\tilde{x}=B^{-1}b$, the only feasible change, given the constraints $x\geq 0$ in (1), is to increase some component(s) of y, while keeping $\tilde{x}=B^{-1}b-B^{-1}Dy\geq 0$.
Hence, it is obvious that the *Simplex criterion* $d'-\tilde{c}'B^{-1}D\geq 0$ is sufficient for the optimality of that feasible basic solution. If the feasible basic solution is

nondegenerated, i.e. $B^{-1}b > 0$, then this condition is obviously necessary for optimality too. But also if *degenerated* feasible basic solutions occur (i.e. some basic variables become equal to zero), one can prove

Theorem 6. *The linear program* (1) *has a solution if and only if there is a feasible basis which satisfies the Simplex criterion.*

Now the Simplex method works as follows: Start with a feasible basis B. If the Simplex criterion is satisfied, we have an optimal feasible basic solution with an optimal value $\tilde{c}'B^{-1}b$. Otherwise, increase a y_j, for which $(d' - \tilde{c}'B^{-1}D)_j < 0$, until an $\tilde{x}_i$ vanishes. If $\tilde{x}_i$ is the first to vanish, exchange the i-th column of B and the j-th column of D, i.e. let y_j become a basic variable and $\tilde{x}_i$ a nonbasic variable. After rearranging $\tilde{c}$ and d correspondingly, restart from the beginning. The arithmetic operation — necessary for this exchange of basic and nonbasic variables — is called *pivoting.*

As long as only nondegenerated feasible basic solutions occur, it is obvious that in (3) the value of the objective function is strictly decreased at each step. Thus, the procedure must then be finite, since there are only finitely many feasible basic solutions. There are additional rules to avoid cycling, in case degeneracy occurs.

Some special methods have been developed making use of special data structures in the matrix A, one of which is the so-called *decomposition structure.* Here A looks like

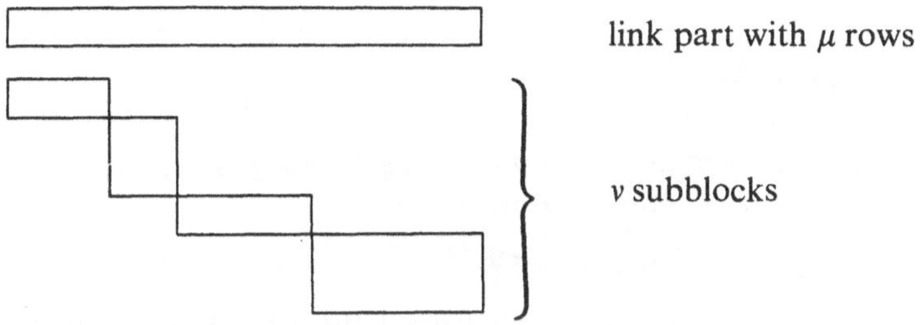

where all elements outside the rectangles are equal to zero. *Decomposition methods* take advantage of this structure — especially for large-scale problems — by pivoting in intermediate steps only on the subblocks of A, whereas in the main pivot steps one is concerned with a matrix of $\mu + \nu$ rows (corresponding to the so-called masterprogram). Without going into details we can say that this procedure is essentially based on Th. 3.

In certain practical situations it is of interest to know what happens to the solution or the optimal value of a linear program if some of the elements of A, b, c vary. This kind of problem leads to *parametric linear programming.* For later purposes we might mention the following special result:

Theorem 7. *Let* $T \subset \mathbb{R}^m$ *be a convex (polyhedral) set. Suppose that* $\{x \mid Ax = t, x \geq 0\} \neq \emptyset$ *for all* $t \in T$ *and that* $\gamma(t) = \min \{c'x \mid Ax = t, x \geq 0\}$ *is finite for at least one* $t \in T$. *Then* $\gamma(t)$ *is finite for all* $t \in T$ *and* $\gamma(t)$ *is a piecewise linear, convex function on* T *(i.e. for arbitrary* $t^1 \in T$, $t^2 \in T$ *and* $\lambda \in (0,1)$ *we have* $\gamma(\lambda t^1 + (1-\lambda)t^2) \leq \lambda\gamma(t^1) + (1-\lambda)\gamma(t^2)$).

This assertion can be proved by Th. 4 and Th. 6.

Finally, we have to mention the *duality* in linear programming. To every so-called *primal* linear program,

(4) $$\min\{c'x \mid x \in \mathbb{R}^n, Ax=b, x \geq 0\},$$

we may express its so-called *dual* program as

(5) $$\max\{b'u \mid u \in \mathbb{R}^m, A'u \leq c\}.$$

Theorem 8. *Suppose that* (4) *and* (5) *have feasible solutions* $\hat{x}$ *and* $\hat{u}$, *respectively. Then*
a) $b'\hat{u} \leq c'\hat{x}$, *and*
b) (4) *and* (5) *have a solution.*

Assertion a) follows from the feasibility of $\hat{x}$ and $\hat{u}$, and b) follows from Th. 2.

Theorem 9. *(Duality Theorem).* (4) *has a solution* x^* *if and only if* (5) *has a solution* u^*. *Then* $c'x^* = b'u^*$.

For further details see

George B. Dantzig: Linear Programming and Extensions. Princeton University Press Princeton.

2. Nonlinear Programming

A *nonlinear program* is a mathematical programming problem which is not a linear program, i.e. a nonlinear program is of the form

(6) $$\min\{f(x) \mid x \in \mathfrak{B}\}$$

where the objective function $f(x)$ is some real valued function defined on the feasible set $\mathfrak{B} \subset \mathbb{R}^n$. However, in this generality problem (6) cannot be handled. One of the reasons is the fact that in general a local minimum of (6) is not a global one. We may overcome this difficulty by assuming that (6) is a convex program. Problem (6) is a *convex program* if the feasible *set* is *convex*, i.e. if $x \in \mathfrak{B}$ and $y \in \mathfrak{B}$, then $\lambda x + (1-\lambda)y \in \mathfrak{B}$ for all $\lambda \in (0,1)$, and if f is a *convex function*, i.e. for $x \in \mathfrak{B}$ and $y \in \mathfrak{B}$ we have $f(\lambda x + (1-\lambda)y) \leq \lambda f(x) + (1-\lambda)f(y)$ for all $\lambda \in (0,1)$. From the convexity of a function f we may conclude immediately the following statements:

Theorem 10. *Let* $\mathfrak{B} \subset \mathbb{R}^n$ *be a convex set and* $f: \mathfrak{B} \to \mathbb{R}$ *be a convex function. Then* f *is continuous in the interior of* $\mathfrak{B}$.

Theorem 11. *Let* $f: \mathfrak{B} \to \mathbb{R}$ *be convex and differentiable at* $x^1 \in \mathfrak{B}$. *Then* $f(x^2) - f(x^1) \geq (x^2 - x^1)' \operatorname{grad} f(x^1)$ *for all* $x^2 \in \mathfrak{B}$.

Theorem 12. *Let* $f: \mathfrak{B} \to \mathbb{R}$ *be convex and twice differentiable. Then the matrix* $(\partial f / \partial x_i \partial x_j)$ *is positive semidefinite on* $\mathfrak{B}$.

We have to mention that a function f is called *concave*, if $-f$ is convex.

Usually the feasible set $\mathfrak{B}$ of a convex program is determined by constraints, i.e. $\mathfrak{B}=\{x|g_i(x)\leq 0,\ i=1,\ldots,k\}$, where $g_i(x)$, $i=1,\ldots,k$, are convex functions. For the following, it is meaningful to distinguish between linear and nonlinear constraints, so that $\mathfrak{B}=\{x|g_i(x)\leq 0, i=1,\ldots,k; h_i(x)=0, i=1,\ldots,m\}$, where $h_i(x)=\sum_{j=1}^{n} a_{ij}x_j-b_i$. Hence, we rewrite (6) in the following form:

$$\begin{aligned}&\min f(x)\\ &\text{subject to}\quad g(x)\leq 0\\ &\qquad\qquad\quad h(x)=0,\end{aligned}\tag{7}$$

where g and h are vector-valued functions $g:\mathbb{R}^n\to\mathbb{R}^k$ and $h:\mathbb{R}^n\to\mathbb{R}^m$, whose components have the above-mentioned properties. To get an existence theorem, we require a *regularity condition*, for example the following:

$$\exists \hat{x}\in\mathbb{R}^n: g(\hat{x})<0,\qquad h(\hat{x})=0.\tag{8}$$

Then we may state

Theorem 13. (Kuhn-Tucker Theorem). *Given the regularity condition* (8), *the convex program* (7) *has a solution if and only if there exist* $\bar{x}\in\mathbb{R}^n$, $\bar{u}\in\mathbb{R}^k$, $\bar{u}\geq 0$, *and* $\bar{v}\in\mathbb{R}^m$ *such that*

$$f(\bar{x})+u'g(\bar{x})+v'h(\bar{x})\leq f(\bar{x})+\bar{u}'g(\bar{x})+\bar{v}'h(\bar{x})\leq f(x)+\bar{u}'g(x)+\bar{v}'h(x)$$

for all $x\in\mathbb{R}^n$, *all* $v\in\mathbb{R}^m$ *and all* $u\in\mathbb{R}^k$ *such that* $u\geq 0$. *Then* $\bar{x}$ *is a solution of* (7).

If we start with another standard formulation of convex programs, namely

$$\begin{aligned}&\min f(x)\\ &\text{subject to}\quad g(x)\leq 0\\ &\qquad\qquad\quad x\geq 0\end{aligned}\tag{9}$$

then — given the corresponding regularity condition — we have, according to Th. 13, to prove the existence of a *saddlepoint* $(\bar{x}',\bar{u}')'\geq 0$ of

$$\begin{aligned}&\Phi(x,u)=f(x)+u'g(x)\quad\text{on}\quad x\geq 0, u\geq 0,\quad\text{i.e.}\\ &\Phi(\bar{x},u)\leq\Phi(\bar{x},\bar{u})\leq\Phi(x,\bar{u})\quad\text{for all}\quad x\geq 0\quad\text{and all}\quad u\geq 0.\end{aligned}$$

Local conditions for such a saddlepoint are given by

Theorem 14. *Let* $\varphi(x,u)$ *be convex in* x *and concave in* u *and continuously differentiable. Then* $(\bar{x}',\bar{u}')'$ *is a saddlepoint of* φ, i.e. $\bar{x}\geq 0, \bar{u}\geq 0$ *and* $\varphi(\bar{x},u)\leq\varphi(\bar{x},\bar{u})\leq\varphi(x,\bar{u})$ *for all* $x\geq 0, u\geq 0$, *if and only if for* $i=1,\ldots,n$ *and* $j=1,\ldots,k$

$$\begin{aligned}&\frac{\partial\varphi(\bar{x},\bar{u})}{\partial x_i}\geq 0 &&\frac{\partial\varphi(\bar{x},\bar{u})}{\partial u_j}\leq 0\\ &\bar{x}_i\cdot\frac{\partial\varphi(\bar{x},\bar{u})}{\partial x_i}=0 &&\bar{u}_j\frac{\partial\varphi(\bar{x},\bar{u})}{\partial u_j}=0\\ &\bar{x}_i\geq 0 &&\bar{u}_j\geq 0.\end{aligned}$$

The application of this theorem to problem (9) with $\varphi(x,u)=f(x)+u'g(x)$ yields the *local Kuhn-Tucker conditions* for a solution of (9):

$$(10)\qquad \begin{array}{rr} \operatorname{grad} f(\bar{x}) + (\bar{u}'G(\bar{x}))' \geq 0 & g(\bar{x}) \leq 0 \\ \bar{x}'[\operatorname{grad} f(\bar{x}) + (\bar{u}'G(\bar{x}))'] = 0 & \bar{u}'g(\bar{x}) = 0 \\ \bar{x} \geq 0 & \bar{u} \geq 0 \end{array}$$

where $G(x)$ is a matrix with the elements $G_{ij}(\bar{x}) = \dfrac{\partial g_i(\bar{x})}{\partial x_j}$.

In particular, for a so-called *quadratic program*

$$(11)\qquad \begin{array}{rl} & \min\left[c'x + \frac{1}{2}x'Qx\right] \\ \text{subject to} & Ax \geq b \\ & x \geq 0, \end{array}$$

where Q is a real symmetric positive semidefinite $(n \times n)$-matrix (see Th. 12), the local Kuhn-Tucker conditions are:

$$(12)\qquad \begin{array}{rr} c + Q\bar{x} - A'\bar{u} \geq 0 & b - A\bar{x} \leq 0 \\ \bar{x}'(c + Q\bar{x} - A'\bar{u}) = 0 & \bar{u}'(b - A\bar{x}) = 0 \\ \bar{x} \geq 0 & \bar{u} \geq 0. \end{array}$$

There are different approaches for solving convex programs. Besides linearization methods, which approximate the original program by linear programs, there are gradient methods, applied to the original problem, and so-called complementarity methods, applied to the Kuhn-Tucker conditions. In the special case of quadratic programming, gradient methods as well as complementarity methods are available, which simply require pivoting — as in the simplex method — under additional rules for choosing the pivot elements.

For further details see

G. Hadley: Nonlinear and Dynamic Programming, Addison-Wesley Publ. Company, Inc., Reading — Palo Alto — London.

3. Measure Theory and Probability Theory

One of the basic concepts in measure theory is that of a *measurable space* $(R, \mathfrak{A})$, where R is some nonempty set (called space) and $\mathfrak{A}$ is a σ-algebra on R. $\mathfrak{A}$ is a *σ-algebra*, if it is a nonemty class of subsets of R, which is closed under the formation of complements and countable unions, i.e. if $A \in \mathfrak{A}$, then $\bar{A} = R - A \in \mathfrak{A}$, and if $A_i \in \mathfrak{A}$, $i = 1, 2, 3, \ldots\ldots$, then $\bigcup_{i=1}^{\infty} A_i \in \mathfrak{A}$.

From this definition follows immediately

Theorem 15. *Let* $(R, \mathfrak{A})$ *be a measurable space. If* $A_i \in \mathfrak{A}$, $i = 1, 2, 3, \ldots$, *then* $A_1 - A_2 \in \mathfrak{A}, \bigcap_{i=1}^{\infty} A_i \in \mathfrak{A}$, $\quad \emptyset \in \mathfrak{A}$ *and* $R \in \mathfrak{A}$.

Sets belonging to $\mathfrak{A}$ are called *measurable.*

If in R some nonempty class $\mathfrak{C}$ of subsets is given, we may define in a unique way a "smallest" σ-algebra containing $\mathfrak{C}$, more precisely:

Theorem 16. *If $\mathfrak{E}$ is any nonempty class of subsets of R, then there exists a unique σ-algebra $\mathfrak{A}$ such that $\mathfrak{E} \subset \mathfrak{A}$ and such that if $\mathfrak{C}$ is any other σ-algebra containing $\mathfrak{E}$ then $\mathfrak{A} \subset \mathfrak{C}$.*

This smallest σ-algebra $\mathfrak{A}$ containing $\mathfrak{E}$ is called the *σ-algebra generated by* $\mathfrak{E}$. One of the most important σ-algebras in applications is the *Borel algebra* $\mathfrak{B}$ in $\mathbb{R}^n$, which is the σ-algebra generated by $\mathfrak{E} = \{A_t \mid A_t = \{x \mid x \in \mathbb{R}^n, x \leq t, t \in T\}\}$, where T is the set of all n-tuples of rationals. It is obvious, that for $a \in \mathbb{R}^n$, $b \in \mathbb{R}^n$ sets of the type $\{x \mid x \leq b\}$, $\{x \mid x < b\}$, $\{x \mid x \geq a\}$, $\{x \mid x > a\}$, $\{x \mid a < x \leq b\}$ etc. are Borel sets. From the properties of the natural topology in $\mathbb{R}^n$ follows

Theorem 17. *Every open set in $\mathbb{R}^n$ is a Borel set.*

By the expression *extended real* we indicate that either a real value or $\pm\infty$ may occur. Should extended real components be allowed in $\mathbb{R}^n$, we indicate this by the symbol $\overline{\mathbb{R}}^n$. The Borel algebra $\mathfrak{B}$ is then extended accordingly.
Let $(R, \mathfrak{A})$ and $(S, \mathfrak{C})$ be two measurable spaces.
A mapping $T: R \to S$ is called a *measurable transformation* if $T^{-1}[C] = \{r \mid r \in R; Tr \in C\} \in \mathfrak{A}$ for all $C \in \mathfrak{C}$. If $S = \overline{\mathbb{R}}$ and $\mathfrak{C}$ is the Borel algebra on $\overline{\mathbb{R}}$, T is called a *measurable function.* If, in particular, $R = \mathbb{R}^m$, $S = \overline{\mathbb{R}}^n$ and $\mathfrak{A}$ and $\mathfrak{C}$ are the corresponding Borel algebras, we use the term *Borel measurable transformation.* If moreover $S = \overline{\mathbb{R}}$ and $\mathfrak{C}$ is the Borel algebra on $\overline{\mathbb{R}}$, we speak of a *Borel measurable function.*

Theorem 18. *Let $(R, \mathfrak{A})$, $(S, \mathfrak{C})$, $(U, \mathfrak{D})$ be measurable spaces and $T_1: R \to S$, $T_2: S \to U$ be measurable transformations. Then $T_2 \circ T_1: R \to U$, defined by $T_2 \circ T_1(r) = T_2(T_1(r))$ for $r \in R$, is a measurable transformation.*

Theorem 19. *Let $(R, \mathfrak{A})$ be a measurable space and $f_i: R \to \overline{\mathbb{R}}$, $i = 1, 2, \ldots$, be extended real valued measurable functions. Then $|f_1|, f_1 + f_2, f_1 \cdot f_2$, $\inf_i f_i$ and $\sup_i f_i$ are measurable functions.*

From Th. 17 follows

Theorem 20. *If $T: \mathbb{R}^m \to \mathbb{R}^n$ is continuous, then T is Borel measurable.*

A *measure* on a σ-algebra $\mathfrak{A}$ is a function $\mu: \mathfrak{A} \to \overline{\mathbb{R}}$, with the properties: $\mu(\emptyset) = 0$, $\mu(A) \geq 0$ for all $A \in \mathfrak{A}$ and

$$\mu\left(\bigcup_{i=1}^{\infty} A_i\right) = \sum_{i=1}^{\infty} \mu(A_i)$$

for every countable class of disjoint sets $A_i \in \mathfrak{A}$. If $\mathfrak{A}$ is a σ-algebra on the space R and μ is a measure on $\mathfrak{A}$, then $(R, \mathfrak{A}, \mu)$ is called a *measure space.* A *measure* μ is called *σ-finite,* if there is a countable class of sets $A_i \in \mathfrak{A}$, such that $\mu(A_i) < \infty$ for $i = 1, 2, \ldots$, and $\bigcup_{i=1}^{\infty} A_i = R$. A *measure* μ is *finite,* if $\mu(R) < \infty$. An important example of a σ-finite measure is the *Lebesgue measure* in $\mathbb{R}^n$, which is uniquely determined on the Borel algebra by requiring

$\mu(\{x \mid a < x \leq b\}) = \prod_{i=1}^{n} (b_i - a_i)$ for all $a \in \mathbb{R}^n$, $b \in \mathbb{R}^n$ such that $a < b$.

With respect to a measure space $(R, \mathfrak{A}, \mu)$, a proposition is said to be true *almost everywhere (a.e.)* if the proposition is true for every element of R except at most a measurable set N of elements with $\mu(N)=0$. Hence, a sequence $\{f_n\}$ of measurable functions defined on a measure space $(R, \mathfrak{A}, \mu)$ *converges* to f *a.e.* if there is a $N \in \mathfrak{A}$ such that $\lim_{n\to\infty} f_n(r) = f(r)$ for all $r \in R-N$ and $\mu(N)=0$. The sequence $\{f_n\}$ *converges in measure* to f, if $\lim_{n\to\infty} \mu(\{r \,|\, |f_n(r)-f(r)| \geq \varepsilon\})=0$ for every $\varepsilon > 0$.

To introduce integration we need simple functions. A *simple function* on a measure space $(R, \mathfrak{A}, \mu)$ is a measurable function f, which attains a finite number of different real values γ_i, $i=1,\ldots,k$. If for $A_i = \{r \,|\, f(r)=\gamma_i\}$, $i=1,\ldots,k$, it is true that $\mu(A_i) < \infty$ whenever $\gamma_i \neq 0$, then f is called an *integrable simple function*, and the *integral* is defined as

$$\int f \, d\mu = \sum_{i=1}^{k} \gamma_i \mu(A_i),$$

where for $\gamma_{i_0}=0$ and $\mu(A_{i_0})=\infty$ the product $\gamma_{i_0} \cdot \mu(A_{i_0})=0$ by definition. A sequence $\{f_n\}$ of integrable simple functions is called *mean fundamental*, if $\int |f_n - f_m| \, d\mu$ tends to zero whenever n and m tend to infinity. Now a *measurable function* f on a measure space $(R, \mathfrak{A}, \mu)$ is called *integrable* if there is a sequence $\{f_n\}$ of integrable simple functions which is mean fundamental and converges in measure to f. Then the *integral* is defined as $\int f \, d\mu = \lim_{n\to\infty} \int f_n \, d\mu$.

Theorem 21. a) *A measurable function f on $(R, \mathfrak{A}, \mu)$ is integrable if and only if its absolute value is integrable; and $|\int f \, d\mu| \leq \int |f| \, d\mu$.*

b) *Let f, g be integrable functions on $(R, \mathfrak{A}, \mu)$ and α, β be real constants. Then $\alpha f + \beta g$ is integrable and $\int (\alpha f + \beta g) \, d\mu = \alpha \int f \, d\mu + \beta \int g \, d\mu$.*

Theorem 22 (Legesgue). *If $\{f_n\}$ is a sequence of integrable functions converging in measure* (*or* a.e.) *to f, and if g is an integrable function such that*

$$|f_n(r)| \leq |g(r)| \text{ a.e.}, \; n=1,2,\ldots,$$

then f is integrable and

$$\lim_{n\to\infty} \int |f - f_n| \, d\mu = 0.$$

Theorem 23 (Hölder's Inequality). *Let p and q be real numbers greater than* 1 *such that $\frac{1}{p}+\frac{1}{q}=1$ and assume that f^p and g^q are integrable functions on $(R, \mathfrak{A}, \mu)$. Then the product fg is integrable and*

$$\int |fg| \, d\mu \leq \left(\int |f|^p \, d\mu\right)^{\frac{1}{p}} \left(\int |g|^q \, d\mu\right)^{\frac{1}{q}}.$$

For $p=q=2$ Hölder's inequality is called *Schwarz's Inequality*.

Let $(R, \mathfrak{A})$ be a measurable space and ν, μ be two measures defined on $\mathfrak{A}$. The measure ν is called *absolutely continuous* with respect to μ, $\nu \ll \mu$, if $\mu(A)=0$ implies $\nu(A)=0$.

Theorem 24 (Radon-Nikodym). *If ν and μ are σ-finite measures, then $\nu \ll \mu$ if and only if there is an integrable function f such that $\nu(A) = \int_A f \, d\mu$ for every measurable set A. f is* a.e. *uniquely determined.*
Here $\int_A f \, d\mu = \int \chi_A f \, d\mu$, where $\chi_A(r)=1$ if $r \in A$ and $\chi_A(r)=0$ if $r \notin A$.

Finally we have to mention product spaces. Let $(R, \mathfrak{A}, \mu)$ and $(S, \mathfrak{C}, \nu)$ be σ-finite measure spaces. Then the *product space* is denoted as $(R \times S, \mathfrak{A} \times \mathfrak{C}, \mu \times \nu)$, where $R \times S$ is the Cartesian product of R and S, $\mathfrak{A} \times \mathfrak{C}$ is the σ-algebra generated by the class of all Cartesian products $A \times C$, where $A \in \mathfrak{A}$ and $C \in \mathfrak{C}$, and $\mu \times \nu$ is the *product measure* on $\mathfrak{A} \times \mathfrak{C}$ uniquely determined by $(\mu \times \nu)(A \times C) = \mu(A) \cdot \nu(C)$ for all $A \in \mathfrak{A}$ and $C \in \mathfrak{C}$. If for example $\mathfrak{B}^k$ is the Borel algebra on $\mathbb{R}^k$ and μ_k is the corresponding Lebesgue measure, then $(\mathbb{R}^{m+n}, \mathfrak{B}^{m+n}, \mu_{m+n}) = (\mathbb{R}^m \times \mathbb{R}^n, \mathfrak{B}^m \times \mathfrak{B}^n, \mu_m \times \mu_n)$.

If $D \subset R \times S$, then a *section* of D determined by $r \in R$, or a *r-section* of D, is the set $D_r = \{s \mid (r, s) \in D\}$.

Theorem 25. *In the product space* $(R \times S, \mathfrak{A} \times \mathfrak{C}, \mu \times \nu)$, *a measurable set* $D \subset R \times S$ *has measure zero if and only if almost every r-section (or almost every s-section) of D has measure zero.*

With respect to integration we need the important

Theorem 26 (Fubini's Theorem). *Let f be an integrable function on the product space* $(R \times S, \mathfrak{A} \times \mathfrak{C}, \mu \times \nu)$.
Then

$$\int f d(\mu \times \nu) = \int \left(\int f d\mu \right) d\nu = \int \left(\int f d\nu \right) d\mu.$$

Probability theory may be understood as a special area of measure theory. A *probability space* is a finite measure space $(\Omega, \mathfrak{F}, P)$ for which $P(\Omega) = 1$. The measurable sets (i.e. the elements of $\mathfrak{F}$) are called *events* and P is called a *probability measure.* Instead of a.e. we use the phrase *almost sure.* A measurable transformation $x: \Omega \rightarrow \mathbb{R}^n$ (where the σ-algebra on $\mathbb{R}^n$ is always the Borel algebra) is called a (n-dimensional) *random vector.* A one-dimensional random vector is a *random variable.* Observe that every component of a random vector is itself a random variable. A random vector x defines a probability measure $\tilde{P}$ on $\mathbb{R}^n$ in a natural way by $\tilde{P}(B) = P(x^{-1}[B])$ for all Borel sets B. $\tilde{P}$ is uniquely determined on $\mathfrak{B}^n$ by the *distribution function* of x: $F_x(t) = \tilde{P}(\{\xi \mid \xi \in \mathbb{R}^n, \xi \leq t\})$ for all $t \in \mathbb{R}^n$. If $\tilde{P}$ is absolutely continuous with respect to the Lebesgue measure μ_n, by the Radon-Nikodym theorem there is a *probability density function* $f_x(\tau)$ defined on $\mathbb{R}^n$ such that $\tilde{P}(B) = \int_B f_x(\tau) d\mu_n$ for all $B \in \mathfrak{B}^n$.

The *expectation* Ex of a random vector x is the vector of the integrals of the components of x. For simplicity we write

$$Ex = \left(\int_\Omega x_1 dP, \int_\Omega x_2 dP, \ldots, \int_\Omega x_n dP \right)' = \int_\Omega x dP.$$

Hence we have

$$Ex = \int_\Omega x dP = \int_{R^n} \xi d\tilde{P} = \int_{R^n} \xi dF_x(\xi)$$

where the last expression is the so-called Lebesgue-Stieltjes integral. If x has a probability density function, we may also write $Ex = \int_{R^n} \xi f_x(\xi) d\mu_n = \int_{R^n} \xi f_x(\xi) d\xi$, where $d\mu_n$ and $d\xi$ refer to the Lebesgue measure on $\mathfrak{B}^n$. We call k random variables x_i, $i = 1, \ldots, k$ *stochastically independent,* if

$$P\left(\bigcap_{i=1}^{k} \{\omega \mid x_i(\omega) \in B_i\} \right) = \prod_{i=1}^{k} P(\{\omega \mid x_i(\omega) \in B_i\})$$

for all Borel sets B_i in $\mathbb{R}$. There is an obvious connection between stochastic independence and product measures. Let $\tilde{P}_i$ be the probability measure on $\mathbb{R}$ corresponding to the random variable x_i, $i=1,\ldots,k$, and let $\tilde{P}$ be the probability measure on $\mathbb{R}^k$ corresponding to $x=(x_1, x_2, \ldots, x_k)'$, then stochastic independence of the random variables $x_1, x_2, \ldots, x_k$ is equivalent to $\tilde{P}=\tilde{P}_1 \times \tilde{P}_2 \times \cdots \times \tilde{P}_k$, $F_x(t)=F_{x_1}(t_1)\cdot F_{x_2}(t_2)\cdot \cdots \cdot F_{x_k}(t_k)$ and, if the densities exist, $f_x(\tau)=f_{x_1}(\tau_1)\cdot f_{x_2}(\tau_2)\cdot \cdots \cdot f_{x_k}(\tau_k)$.

From Fubini's Theorem follows

Theorem 27. *Let x_1 and x_2 be stochastically independent and assume that Ex_1, Ex_2 and Ex_1x_2 exist. Then $Ex_1x_2=(Ex_1)(Ex_2)$.*

For further details see

Paul R. Halmos: Measure Theory, D. Van Nostrand Company, Inc., Princeton–Toronto–London–New York.

M. Loève: Probability Theory, D. Van Nostrand Company, Inc., Princeton–Toronto–London–New York.

Chapter I. Introduction

In view of the fact that sometimes there seems to be a terminological confusion, it might be useful to try to explain what *stochastic linear programming* is. There are many practical situations for which — at first glance — linear programs are appropriate models. This is especially the case in production problems with (piecewise) linear production functions and (piecewise) linear cost functions, diet problems with (piecewise) linear cost functions, all other optimal mix problems such as oil-refining, distillation of spirits etc., general network flow problems with (piecewise) linear cost functions, critical path scheduling problems, Hitchcock-type transport problems and so on. It is obvious that these and many similar problems are of great practical importance. For this reason the development of linear programming has been explosive during the last 25 years. Since at the same time there has been an equally remarkable development of the computer technology, linear programming can now be looked on as a standard tool for sloving problems as mentioned above.

Let us see under what conditions the application of linear programming can be justified. If we solve one of the above problems by solving a linear program in one of its standard forms

$$\begin{aligned} \min\ & c'x \\ & Ax=b \\ & x\geq 0 \end{aligned} \tag{1}$$

where $x\in\mathbb{R}^n$, $c\in\mathbb{R}^n$, $b\in\mathbb{R}^m$, we must make sure that our problem not only has the linear structure indicated in the linear program, but also that the coefficients in A, b, c are, at least throughout the planning period, fixed known data. But everybody will agree that this is not true in most practical problems. For example, if the linear program represents a production problem, b is the demand vector, A is the matrix of technological coefficients, c is the vector of costs per unit and x is the vector of factors of production, i.e. x is the input into the production process and shall be determined optimally. It is evident that in many practical situations neither the demand vector nor the technological coefficients nor the cost vector are fixed known data. Then there are three possibilities: Either these data are stochastic variables with known (joint) probability distributions or they are stochastic variables with unknown probability distributions or they are not stochastic variables but simply variables. In all these cases a linear programming model does not make sense. At this point we can explain what the subject of this book is:

Stochastic linear programming (SLP) is concerned with problems arising when some or all *coefficients* of a linear program are *stochastic variables* with *known (joint) probability distribution.*

In this respect many users of linear programming have already been involved in a special procedure of stochastic linear programming, namely by replacing the random variables in a linear program by their expectation values or, fairly good estimates of them, and solving the resulting linear program. The following numerical example shows that this procedure is not feasible in all practical situations. Suppose that the problem is

$$\begin{aligned} \min\ & x_1+x_2 \\ & ax_1+x_2\geq 7 \\ & bx_1+x_2\geq 4 \\ & x_1\geq 0,\ x_2\geq 0 \end{aligned}$$

where (a, b) is a uniformly distributed random vector within the rectangle $\left\{(1\leq\alpha\leq 4), \left(\frac{1}{3}\leq\beta\leq 1\right)\right\}$. Then $E(a, b)=\left(\frac{5}{2}, \frac{2}{3}\right)$, so that the linear program would be

$$\begin{aligned} \min\ & x_1+x_2 \\ & \frac{5}{2}x_1+x_2\geq 7 \\ & \frac{2}{3}x_1+x_2\geq 4 \\ & x_1\geq 0,\ x_2\geq 0, \end{aligned}$$

which yields the unique optimal solution

$$x_1^*=\frac{18}{11},\ x_2^*=\frac{32}{11}.$$

If we ask for the probability of the event, that this solution is feasible with respect to the original problem, we get

$$P\{(a, b)\,|\,ax_1^*+x_2^*\geq 7;\ bx_1^*+x_2^*\geq 4\}=P\left\{(a, b)\,|\,a\geq\frac{5}{2},\ b\geq\frac{2}{3}\right\}=\frac{1}{4}.$$

So this solution is infeasible with probability .75. If we associate with this simple example any practical problems such as diet problems in hospitals or oil refining problems involving such high quality restrictions as for aircraft gasoline, we must agree that in many practical situations the approach chosen above cannot be allowed. And even in cases where human safety is not involved, it seems to be worthwhile to consider the loss of other goods, which may correspond to infeasible solutions. Therefore, one should be careful when using the above procedure in view of the possible practical consequences of infeasibility.

There are essentially two different types of models in SLP situations, namely the so-called "wait and see" and the "here and now" models. "Wait and see" problems are based on the assumption, that the decision maker is able to wait for the realisation of the random variables and to make his decision with complete information on this realization, i.e. if $(\hat{A}, \hat{b}, \hat{c})$ is a realization of the random vector (A, b, c), he has to solve the linear program

$$
(2) \qquad \begin{aligned} \min\ & \hat{c}'x \\ & \hat{A}x = \hat{b} \\ & x \geq 0. \end{aligned}
$$

Typical questions in this case are: What is the expectation of the optimal value of (2), or what are the expectation and the variance of this optimal value and so on. More generally the question is: What is the probability distribution of the optimal value of (2)? A possible interpretation of this *distribution problem* is the following: Suppose that a special production program (with linear structure) may be adapted for any short period to actual realizations of random factor prices, random technological coefficients and random demands. Planning the budget for a long term — i.e. for many short periods — the board of the firm wants to know the amount of cash needed for this production program "in the mean" or "for 95% of the time". More precisely, the board wants to know the expectation, or the 95% percentile, of the probability distribution of this special production program's costs per (short) period.

"Here and now" models are based on the following assumption: A decision on x — or on a "strategy" for x — has to be made in advance or at least without knowledge of the realization of the random variables.

By a "strategy" for x we understand the game theoretical concept of "mixed strategy" within a feasible set X of pure strategies x; or, equivalently, a "strategy" for x is a probability measure P_x on a Borel set $X \subset \mathbb{R}^n$. If we restrict ourselves to probability distributions P_x such that there exists an $\hat{x} \in X$ with $P_x(\{\hat{x}\}) = 1$, we are restricted to pure strategies, i.e. to decisions on x instead of mixed strategies of x's.

The practical interpretation of a strategy is as usual the assumption that the decision maker plays his game very often with — possibly different — x's resulting from a Monte-Carlo simulation of the chosen probability distribution P_x.

To understand the philosophy of "here and now" models, it seems to be necessary to start at the very beginning. Our first observation is that in a linear program some or all coefficients are random variables with a joint probability distribution. This implies — by definition of random vectors — the existence of a probability space $(\Omega, \mathfrak{F}, P_\omega)$ such that $\{A(\omega), b(\omega), c(\omega)\}$ is a measurable transformation on Ω into $\mathbb{R}^{m \times n + m + n}$. Our general assumption for SLP is that we know P_ω. A further very important assumption is that a decision on x — or on a mixed strategy P_x — does not influence P_ω. More precisely, the events in Ω — i.e. the elements of $\mathfrak{F}$ — and the events in X, i.e. the Borel sets in X — are stochastically independent; or equivalently, the probability measure of the product space $X \times \Omega$ is the product measure $P_x \times P_\omega$. It should be pointed out very clearly, that this assumption is not at all trivial from the practical point of view. If for example a producer with a large share in the factor market takes very extreme decisions on inputs, it seems very unlikely that these decisions do not influence input prices or quality, which would alter the technological coefficients. On the other hand there are certainly many cases, where the assumption of stochastic independence is quite realistic. Therefore, in most practical cases we must check very seriously whether the influence of the producer's decision on the probability distribution P_ω may be neglected before applying one of the "here and now" models handled in this book.

A decision maker who does not want to choose his strategy at random out of a certain feasible set must have a criterion telling him whether a certain strategy is the "best" one or not. As is well-known, in decision theory there are different concepts of what the "best" may be. One of them is that there is a partial ordering on the set of feasible strategies P_x: then a "best" strategy is not necessarily "better" (with respect to the partial ordering) than, or equivalent to, every other strategy, but there is no other "comparable" and "better" strategy. Since, under a partial ordering, not every pair of strategies need be comparable, it follows that there may be a strategy which is "best" in virtue of not being comparable to any other feasible strategy. This concept has important applications in multi-goal programming.

However, we shall be concerned with a stronger concept of "best" strategy. Let us assume that any two feasible strategies are comparable and that the result of the comparison says that either one strategy is "better" than the other or both strategies are "equivalent". In other words, either we prefer one strategy to the other or we are indifferent. Furthermore, we suppose that the decision maker is consistent in the following sense. When he has preferred a strategy $P_x^{(1)}$ to $P_x^{(2)}$ and also has preferred $P_x^{(2)}$ to $P_x^{(3)}$, then he will prefer $P_x^{(1)}$ to $P_x^{(3)}$. When he is indifferent with respect to $P_x^{(1)}$ and $P_x^{(2)}$, then he also thinks of $P_x^{(2)}$ as equivalent to $P_x^{(1)}$. And when he believes $P_x^{(1)}$ to be equivalent to $P_x^{(2)}$ and $P_x^{(2)}$ to be equivalent to $P_x^{(3)}$, then he will also be indifferent with respect to $P_x^{(1)}$ and $P_x^{(3)}$. To obtain such a preferential scheme we need, on $\mathfrak{P} \times \mathfrak{P}$, a strict linear ordering $\prec$ (the "better" relation) and an equivalence relation $\sim$ (the "indifferent" relation) so that for any $P_x^{(i)} \in \mathfrak{P}$, $i=1, 2, 3$, the following statements hold:

1) One and only one of these relations hold:

$$P_x^{(1)} \prec P_x^{(2)}, \qquad P_x^{(2)} \prec P_x^{(1)}, \qquad P_x^{(1)} \sim P_x^{(2)}.$$

2) If $P_x^{(1)} \prec P_x^{(2)}$ and $P_x^{(2)} \prec P_x^{(3)}$, then $P_x^{(1)} \prec P_x^{(3)}$.
3) $P_x^{(1)} \sim P_x^{(1)}$. And if $P_x^{(1)} \sim P_x^{(2)}$, then $P_x^{(2)} \sim P_x^{(1)}$.
4) If $P_x^{(1)} \sim P_x^{(2)}$ and $P_x^{(2)} \sim P_x^{(3)}$, then $P_x^{(1)} \sim P_x^{(3)}$.

In the economic as well as in the mathematical literature the question on the existence of a real-valued criterion function $f: \mathfrak{P} \to \mathbb{R}$ inducing a given preferential scheme, is discussed in great detail. This is not our aim at this point, because in SLP-situations a decision maker does not first have a preferential scheme on the feasible strategies and then look for an order preserving real-valued function, but, conversely, it seems to be more likely that he has a certain criterion function $f: \mathfrak{P} \to \mathbb{R}$ which yields in a very natural way a preferential scheme on $\mathfrak{P} \times \mathfrak{P}$:

a) $f(P_x^{(1)}) < f(P_x^{(2)}) \Rightarrow P_x^{(1)} \prec P_x^{(2)}$
b) $f(P_x^{(1)}) = f(P_x^{(2)}) \Rightarrow P_x^{(1)} \sim P_x^{(2)}$

One can easily verify that these relations "$\prec$" and "$\sim$" defined by a) and b) fulfil the above conditions 1) to 4). We shall discuss now how to construct a criterion function on $\mathfrak{P}$ in a SLP situation. As we have seen, we shall then have also solved the problem — at least implicitly — of how to get a preferential scheme on $\mathfrak{P} \times \mathfrak{P}$.

The following construction has turned out to be appropriate to SLP-problems. Given a realization of (A, b, c), i.e. for a certain ω, and given a certain decision x which, as outlined above, may also be understood as a realization of a strategy P_x, we are interested in the result $e(\omega,x)=(c'(\omega)\cdot x, A(\omega)x-b(\omega))$, which represents the actual value of the objective function and the vector of "deviations" in the constraints. First we define a loss function L on the set E of possible results $e(\omega, x)$, i.e. $L:E\rightarrow\mathbb{R}$. Obviously $L\{e(\omega, x)\}$ represents the value of loss we assign to both the actual value of $c'(\omega)x$ (for example the actual costs) and the vector of deviations in the constraints. It should be noted that L depends on the actual strategy x *and* the realisation ω. However, what we want to have is a real-valued function on the set $\mathfrak{P}$ of feasible strategies. Therefore we need a transformation Φ which assigns a real-valued function on $\mathfrak{P}$ to any loss function L, i.e. $\Phi:\mathfrak{L}\rightarrow\mathfrak{G}$. where $\mathfrak{L}=\{L|\,L:E\rightarrow\mathbb{R}\}$ and $\mathfrak{G}=\{F|\,F:\mathfrak{P}\rightarrow\mathbb{R}\}$.
Now to define a special "here and now" problem we have to define the set $\mathfrak{P}$ of feasible strategies, to choose a special loss function $L:E\rightarrow\mathbb{R}$ and a special transformation $\Phi:\mathfrak{L}\rightarrow\mathfrak{G}$. The set $\mathfrak{P}$ may, or may not depend on the probability distribution P_ω. Additionally, $\mathfrak{P}$ may be restricted to certain types of probability distributions on X, where, in most practical cases, X is a convex polyhedral set. If $\mathfrak{P}=\{P_x|\,\exists\hat{x}\in X: P_x\{\hat{x}\}=1\}$, we shall write for simplicity $\mathfrak{P}=X$. As we have seen, L depends on $(c'(\omega)x, A(\omega)x-b(\omega))$. Therefore, to define L we — or the decision maker — must know which value or penalty costs have to be assigned to an actual deviation $A(\omega)x-b(\omega)$, and how to take them into account, together with $c'(\omega)x$, which may represent the actual costs. Consequently, Φ represents the influence of all possible losses — with respect to X and Ω — and of P_ω and the feasible P_x on the criterion function. We summarize this construction of the criterion function as follows:
Given a probability space $(\Omega, \mathfrak{F}, P_\omega)$, a measurable transformation $(A(\omega), b(\omega), c(\omega)):\Omega\rightarrow\mathbb{R}^{m\times n+m+n}$ and a Borel set $X\subset\mathbb{R}^n$ of feasible pure strategies, we define

(i) $E=\{(c'(\omega)\cdot x, A(\omega)x-b(\omega))|\,\omega\in\Omega, x\in X\}$
(ii) $\mathfrak{P}=\{P_x|\,P_x$ is a probability measure on X; possibly further conditions$\}$
(iii) $L:E\rightarrow\mathbb{R}$
(iiii) Φ such that $\Phi L=F:\mathfrak{P}\rightarrow\mathbb{R}$.

The problem then is to minimize F on $\mathfrak{P}$.
To show that this construction is not so artificial as it might seem, we shall conclude this section with some examples.

Suppose that
1) $\mathfrak{P}=X\cap\{x|\,P_\omega(A(\omega)x-b(\omega)\geq 0)\geq\alpha\}, \quad \alpha\in[0,1]$
$L(e(\omega, x))=c'x$
$\Phi L=E_\omega\, L(e(\omega, x))$
The problem resulting from these definitions is

$$\min_{\mathfrak{P}} E_\omega L(e(\omega, x))$$

or equivalently

$$\min E_\omega c'(\omega)x$$

with respect to $P_\omega(A(\omega)x-b(\omega)\geq 0)\geq\alpha$

$$x\in X.$$

This is a so-called *chance constrained programming* problem, which means that the expectation of the original objective function $c'x$ shall be minimized with respect to the constraints that $x\in X$ (for example $x\geq 0$) and that the decision — or pure strategy — x must be feasible in $Ax\geq b$ with probability at least α.

2) We have an entirely different type of problem, if

$$\mathfrak{P}=\{P_x \mid P_x(X)=1\}$$

$$L(e(\omega,x))=c'(\omega)\cdot x+\min\{q'y \mid Wy=b(\omega)-A(\omega)x, \quad y\geq 0\},$$

where W, q may also be deterministic or stochastic, in the sense that they depend on $\omega\in\Omega$,

$$\Phi L=F(E_{P_x\times P_\omega} L^i(e(\omega,x)); \quad i=1,\dots,r).$$

In particular, the case when $F=\mu E_{P_x\times P_\omega}\{L(e(\omega,x))\}+\lambda\cdot\sigma_{P_x\times P_\omega}\{L(e(\omega,x))\}$ with $\lambda\geq 0$ seems to be of practical importance.

This type of problem is called a *two stage problem* of SLP or a SLP *problem with recourse*. The practical meaning of problems with recourse is this:

When we have determined x (or x has been determined as a realization of a strategy P_x) and when we observe a realization of the random vectors $(A(\omega), b(\omega), c(\omega))$, then there may be a deviation from the original constraints, i.e. $A(\omega)x=b(\omega)$ may not be satisfied. Such a deviation causes penalty costs arising from a second stage linear program $\min\{q'y \mid Wy=b(\omega)-A(\omega)x, y\geq 0\}$, which may be understood as an "emergency program", yielding the last possibility of compensating for the deviation from the original constraints. The total costs observed in this situation are the sum of the original costs $c'(\omega)x$ and the penalty costs, and — since they depend on ω and x — these total costs are random. The objective is to determine x or P_x such that some function F of certain moments of the total costs becomes minimal, for example the expected total costs or – if risk aversion is involved – a weighted sum of the expectation and the standard deviation of the total costs.

3) A further possibility is:

$$\mathfrak{P}=X\cap\{x \mid P_\omega(c'(\omega)x\leq\gamma)\geq\alpha\}$$

$$L(e(\omega,x))=\begin{cases}1 \text{ if } A(\omega)x\geq b\\ 0 \text{ otherwise}\end{cases}$$

$$\Phi L=-E_\omega L$$

The problem of minimizing ΦL on $\mathfrak{P}$ can be written as

$$\max_{x\in X} P_\omega(A(\omega)x\geq b(\omega))$$

with respect to $P_\omega(c'(\omega)x\leq\gamma)\geq\alpha$;

this means that the costs $c'(\omega)x$ should not exceed a certain prescribed value γ with probability α and that with respect to this constraint the probability of feasibility, which may be a measure of reliability of the production process, shall be as high as possible. Therefore we call this problem *reliability optimization* in SLP. Though these problems have not yet been investigated satisfactorily, they seem to be of great practical importance in many production processes in which a high quality rather than minimal costs should be achieved. Since the theoretical questions with respect to this problem are essentially the same as in chance constrained programming, we shall not handle them separately.

4) Problems as known in two-person-zero-sum games arise, for example, in the following form:

$$\mathfrak{P} = X$$

$L(e(\omega, x))$ some real valued function

$$\Phi L = \operatorname{ess.sup}_{\omega \in \Omega} L(e(\omega, x))$$

Then we have to determine

$$\min_{x \in X} \operatorname{ess.sup}_{\omega \in \Omega} L(e(\omega, x))$$

However, we shall not discuss these problems here because their general theory is contained in the literature on decision theory and they do not really make use of the information on P_ω, i.e. they just take into account some sets of measure zero with respect to P_ω (in determining $\operatorname{ess.sup}_{\omega \in \Omega} L$).

We shall discuss the distribution problem as well as the problem with recourse and the chance constrained problem. From this discussion the reader may conclude how many problems are still unsolved and to which of them he might direct his efforts.

Whereas the interpretation of the distribution problem should be clear from the description given above, it might be helpful to have an example for chance constrained and two stage programming.

Assume that for producing iron you have two ores with different concentrations a_1 and a_2, and that the iron production should meet essentially the demand b (e.g. per month) of one customer. The costs per unit of the ore may be c_1 and c_2, respectively. According to your production capacity you may not smelt more than d units of ore (per month). If all the data were deterministic, we should have the linear program

$$\begin{aligned} &\min\,(c_1 x + c_2 y) \\ &\text{subject to } a_1 x + a_2 y \geq b \\ &\qquad x + y \leq d \\ &\qquad x \geq 0,\ y \geq 0 \end{aligned}$$

where x and y are the quantities of ores smelted.

Suppose now that only the demand b is random and, for simplicity, uniformly distributed on the interval $1.2 \leq b \leq 1.6$, whereas $a_1 = 0.5$; $a_2 = 0.3$; $c_1 = 2$; $c_2 = 1$ and $d = 4$ are deterministic. Assume further that you do not have a capacity to store iron and that you must order the monthly quantities of ore x and y before knowing the actual iron demand b. However to maintain the good will of your customer, you should be able to deliver the quantity b of iron required with a high probability, e.g. with probability 0.9. This yields the chance constrained program

$$\begin{aligned} \min \quad & [2x+y] \\ \text{subject to } & P(0.5x+0.3y \geq b) \geq 0.9 \\ & x+y \leq 4 \\ & x \geq 0,\ y \geq 0. \end{aligned}$$

In this particular case we get an equivalent linear program, since $P(0.5x+0.3y \geq b) = F_b(0.5x+0.3y)$, where F_b is the distribution function of the random variable b and the constraint $F_b(0.5x+0.3y) \geq 0.9$ is equivalent to $0.5x+0.3y \geq F_b^{-1}[0.9] = 1.56$ according to our assumptions on the probability distribution of b. As we shall see in Ch. IV, chance constrained programs are not equivalent to linear or convex programs in general. In our simple example the assumption of stochastic concentrations a_1 and a_2 would already raise difficulties.

Suppose now a somewhat different situation. You have a contract with your customer to meet exactly his monthly random demand b during the next two years. But you still have to order the ores in advance. If you produce a surplus of iron, you may sell it to other customers at the (low) price $q_1=2$; if you, on the other hand, produce less then b, you must buy yourself the difference at the price $q_2=4$ to fulfil your contract. Hence after knowing the demand b of a particular month, you shall have additional costs $Q(x,y,b)$, following from the recourse program

$$\begin{aligned} Q(x,y,b) = \min & (-2z_1+4z_2) \\ \text{subject to } & z_2-z_1 = b-0.5x-0.3y \\ & z_1 \geq 0, \quad z_2 \geq 0. \end{aligned}$$

Minimizing the total costs (per month) in the mean, i.e. the expected total costs, yields the two stage program

$$\begin{aligned} \min & \ [2x+y+E_b Q(x,y,b)] \\ \text{subject to } & \ x+y \leq 4 \\ & \ x \geq 0, \quad y \geq 0, \end{aligned}$$

where $E_b Q(x,y,b)$ is the expectation of $Q(x,y,b)$ with respect to the distribution of b.

Whether to choose the chance constrained or the recourse model cannot be determined theoretically. This choice depends on the real situation and on the aims of the decision maker.

Chapter II. Distribution Problems

1. The General Case

Let us consider the linear program

$$\text{(1)} \qquad \begin{aligned} &\text{determine } \gamma = \inf c'x \\ &\text{subject to } Ax = b \\ &\qquad\qquad x \geq 0, \end{aligned}$$

where $x \in \mathbb{R}^n$, A is a real $(m \times n)$-matrix, and b and c are real m- and n-vectors respectively. Without loss of generality we assume that $m \leq n$.
If we define $\gamma = +\infty$ whenever (1) has no feasible solution, then $\gamma = \gamma(A, b, c)$ is an extended real-valued function, defined on $\mathbb{R}^{m \times n + m + n}$.
Obviously γ represents the optimal value of (1), if this linear program has a solution.
As we have seen in Ch. I, in an SLP situation there is a known probability space $(\Omega, \mathfrak{F}, P_\omega)$ and a measurable transformation $(A(\omega), b(\omega), c(\omega)): \Omega \to \mathbb{R}^{m \times n + m + n}$ such that $(A, b, c) = (A(\omega), b(\omega), c(\omega))$ is a random vector. Therefore $\gamma(A, b, c)$ becomes $\gamma(\omega) = \gamma(A(\omega), b(\omega), c(\omega))$. Now we may state the general
Distribution problem: Given P_ω, what is the probability distribution of $\gamma(\omega)$?
First of all we have to assure that this problem is mathematically meaningful, i.e. that $\gamma(\omega)$ is a random variable or, equivalently, that $\gamma(\omega): \Omega \to \mathbb{R}$ is measurable. This may be concluded by Th. 0.18 and the following
Theorem 1. *$\gamma(A, b, c): \mathbb{R}^{m \times n + m + n} \to \overline{\mathbb{R}}$ is a Borel measurable extended real-valued function.*

Proof: We shall prove this statement by constructing a countable set of Borel measurable functions $\gamma_{in}: \mathbb{R}^{m \times n + m + n} \to \overline{\mathbb{R}}$ and showing that $\gamma = \sup_n \inf_i \gamma_{in}$. Applying Th. 0.19 we get the desired result.
Let $\mathcal{D}$ be a countable set of vectors $x_i \geq 0$, which is dense in $\{x \mid x \in \mathbb{R}^n, x \geq 0\}$, for example the set of all non-negative n-tuples of rationals.
Define by using as vector-norm the Euclidean norm

$$\mathfrak{A}_{in} = \left\{(A, b, c) \,\middle|\, \|Ax_i - b\| \leq \frac{1}{n}\right\}$$

and

$$\gamma_{in} = \begin{cases} c'x_i & \text{if } (A, b, c) \in \mathfrak{A}_{in} \\ \infty & \text{otherwise} \end{cases}$$

Obviously $\gamma_{in}: \mathbb{R}^{m \times n + m + n} \to \overline{\mathbb{R}}$ is Borel measurable, since the euclidean norm defining the natural topology is continuous and therefore Borel measurable, (see Th. 0.20) which implies $\mathfrak{A}_{in}$ to be Borel measurable sets. Furthermore γ_{in} is continuous in c on $\mathfrak{A}_{in}$.

Let us show, that $\gamma = \sup_n \inf_i \gamma_{in}$, in two steps:

A) $\inf_i \gamma_{in} \leq \gamma$ for $n = 1, 2, \ldots$

We have to distinguish three possible cases:

A1) $\gamma = +\infty$.

In this case $\inf_i \gamma_{in} \leq \gamma$ is trivial.

A2) $-\infty < \gamma < \infty$

Then there exists a vector $x^0 \geq 0$ such that $Ax^0 = b$ and $\gamma = c'x^0$. For an arbitrarily chosen natural number n and any real positive ε let $\delta(\varepsilon, n) = \min\left[\frac{1}{n\|A\|}; \frac{\varepsilon}{\|c\|}\right]$.
Since $\mathcal{D}$ is dense in $\{x \mid x \geq 0\}$, there exists an $x_i \in \mathcal{D}$ such that $\|x^0 - x_i\| \leq \delta(\varepsilon, n)$.
For this x_i we have

$$\|Ax_i - b\| = \|Ax_i - Ax^0\| \leq \|A\| \|x_i - x^0\| \leq \|A\| \cdot \delta(\varepsilon, n) \leq \frac{1}{n}$$

and therefore

$$|\gamma_{in} - \gamma| = |c'x_i - c'x^0| \leq \|c\| \|x_i - x^0\| \leq \|c\| \cdot \delta(\varepsilon, n) \leq \varepsilon.$$

Since we can repeat this procedure for any $\varepsilon > 0$ and for all natural n, we have proved that $\inf_i \gamma_{in} \leq \gamma$ for $n = 1, 2, \ldots$

A3) $\gamma = -\infty$

In this case there exist vectors x^ν, $\nu = 1, 2, 3, \ldots$ such that $x^\nu \geq 0$, $Ax^\nu = b$, $c'x^\nu \leq -\nu$.
Again let

$$\delta(\varepsilon, n) = \min\left[\frac{1}{n\|A\|}; \frac{\varepsilon}{\|c\|}\right].$$

Then to each x^ν there exists a $x_{i_\nu} \in \mathcal{D}$ which satisfies

$$\|x^\nu - x_{i_\nu}\| \leq \delta(\varepsilon, n)$$

Again we have

$$\|Ax_{i_\nu} - b\| \leq \frac{1}{n}$$

and

$$|\gamma_{i_\nu n} - c'x^\nu| \leq \varepsilon$$

and thereby

$$\gamma_{i_\nu n} \leq -\nu + \varepsilon$$

which yields $\inf_i \gamma_{in} = -\infty$ for $n = 1, 2, 3, \ldots$

B) $\sup_n \inf_i \gamma_{in} = \gamma$

B1) $\gamma = -\infty$

From A3) we know that $\inf_i \gamma_{in} = -\infty$ for $n = 1, 2, 3, \ldots$.
Therefore the statement $\sup_n \inf_i \gamma_{in} = \gamma$ is trivial.

B2) $\gamma=\infty$

In this case $\{x \mid Ax=b, x\geq 0\}=\emptyset$ and therefore $\varrho=\inf_{x\geq 0}\|Ax-b\|>0$. This implies

$(A,b,c)\notin\mathfrak{A}_{in}$ for all $n>\frac{1}{\varrho}$ and for all $i:x_i\in\mathcal{D}$.

For all these i and n we have

$\gamma_{in}=\infty$ and consequently $\sup_n\inf_i\gamma_{in}=\infty=\gamma$.

B3) $-\infty<\gamma<\infty$

In this situation there exists an optimal basic solution. Without loss of generality, we may assume that A has rank m. Let B be the optimal basis, i.e. an $(m\times m)$-submatrix of A, and A_N the matrix of non-basic columns of A. Let $\tilde{c}$ be the vector of the components of c belonging to the basic variables and c_N the vector of the remaining components of c. Finally, let $\tilde{x}$ be the vector of basic variables and y the vector of non-basic variables. Then the following relations must hold:

$$\begin{aligned} &B^{-1}b\geq 0 && \text{(feasibility)}\\ &\gamma=\tilde{c}'B^{-1}b &&\\ &c'_N-\tilde{c}'B^{-1}A_N\geq 0 && \text{(optimality).}\end{aligned}$$

If $\sup_n\inf_i\gamma_{in}<\gamma-\varepsilon$ for some $\varepsilon>0$, then for every n there exists a i_n such that $\gamma_{i_n n}<\gamma-\frac{\varepsilon}{2}$. For simplicity we suppose now that for every n there exists $x_{i_n}\in\mathcal{D}$ such that $\|A_i x_{i_n}-b\|\leq\frac{1}{n}$ and $\gamma_{i_n n}=c'x_{i_n}<\gamma$. (Otherwise there would exist an n such that $\gamma_{in}\geq\gamma$ for all i, and by A2) we should conclude that $\sup_n\inf_i\gamma_{in}=\gamma$). Then $x_{i_n}\geq 0$ and $Ax_{i_n}=b+d_n$, where $\|d_n\|\leq\frac{1}{n}$.

Consider the linear program

$$\text{(2)}\qquad \begin{aligned} &\tau_n=\inf c'x\\ &\text{subject to } Ax=b+d_n\\ &\qquad\qquad\quad x\geq 0.\end{aligned}$$

Obviously $\tau_n\leq\gamma_{i_n n}<\gamma$, since x_{i_n} is feasible in (2). On the other hand τ_n must be finite, because $\tau_n=-\infty$ would imply the existence of a vector $w\geq 0, Aw=0$, $c'w<0$ (see Th. 0.4), which contradicts our assumption that (1) had a finite optimal value γ. With respect to the basis B the basic and non-basic variables $\tilde{x}$ and y of any optimal solution of (2) must satisfy the equation

$$\tilde{x}=B^{-1}b+B^{-1}d_n-B^{-1}A_N y.$$

Since

$$c'_N-\tilde{c}'B^{-1}A_N\geq 0 \quad\text{and}\quad y\geq 0,$$

we get

$$\begin{aligned}\tau_n&=\tilde{c}'B^{-1}b+\tilde{c}'B^{-1}d_n+(c'_N-\tilde{c}'B^{-1}A_N)y\\ &\geq\gamma+\tilde{c}'B^{-1}d_n\end{aligned}$$

and therefore

$$0<\gamma-\gamma_{i_n n}\leq\gamma-\tau_n\leq|\tilde{c}'B^{-1}d_n|\leq\|\tilde{c}\|\cdot\|B^{-1}\|\cdot\frac{1}{n},$$

and this implies again that $\sup_n\inf_i\gamma_{in}=\gamma$.

$\{\gamma_{in}\}$ is a countable class of Borel measurable functions. By Th. 0.19, $\{\inf_i \gamma_{in}\}$ is a countable class of Borel measurable functions and, again by Th. 0.19, $\gamma = \sup_n \inf_i \gamma_{in}$ is therefore Borel measurable. □

If we consider the optimal solution instead of the optimal value of (1), the conjecture that this optimal solution is also a Borel measurable transformation must be false, since the optimal solution is in general not uniquely determined. However, we may prove the following

Theorem 2. *There is a Borel measurable transformation* $\hat{x}: \mathbb{R}^{m\times n+m+n} \to \overline{\mathbb{R}}^n$ *which coincides with an optimal solution of* (1), *whenever* (1) *has a solution.*

Proof. Let $\mathscr{R} = \mathbb{R}^{m\times n+m+n}$

$\mathfrak{M} = \{(A,b,c) \mid (1)$ has no feasible solution$\}$

$\mathfrak{Z} = \{(A,b,c) \mid (1)$ has feasible solutions and rank $(A) = m\}$

$\mathfrak{N} = \{(A,b,c) \mid (1)$ has feasible solutions and rank $(A) < m\}$

Obviously $\mathfrak{M}$, $\mathfrak{N}$, $\mathfrak{Z}$ are disjoint sets and

$$\mathscr{R} = \mathfrak{M} \cup \mathfrak{N} \cup \mathfrak{Z}.$$

First we show, that $\mathfrak{M}$, $\mathfrak{N}$, $\mathfrak{Z}$ are Borel measurable.

a) $\mathfrak{M}$ is a Borel measurable subset of $\mathscr{R}$.

As we know from the proof of Th. 1, $\mathfrak{M} = \{(A,b,c) \mid \inf_{x\geq 0} \|Ax-b\| > 0\}$.

Let $\mathscr{D} = \{x_i \mid i = 1,2,\dots\}$ be a countable dense subset of $\{x \mid x \in \mathbb{R}^n, x \geq 0\}$.
Then

$$\inf_{x\geq 0} \|Ax-b\| = \inf_{x_i\in\mathscr{D}} \|Ax_i-b\| \quad \text{for all} \quad (A,b).$$

This follows immediately from the fact, that

$$\inf_{x\geq 0} \|Ax-b\| \leq \inf_{x_i\in\mathscr{D}} \|Ax_i-b\| \quad \text{for all} \quad (A,b)$$

and that simultaneously

$$\inf_{x\geq 0} \|Ax-b\| \geq \inf_{x_i\in\mathscr{D}} \|Ax_i-b\|,$$

which we explain as follows:

For some (A,b) let $\{x^\nu\}$ be a sequence, $x^\nu \geq 0$, such that $\|Ax^\nu - b\|$ tends to $\inf_{x\geq 0} \|Ax-b\|$.

Since $\mathscr{D}$ is dense in $\{x \mid x \geq 0\}$, there exist $x_{i_\nu} \in \mathscr{D}$ with the property that $\|x^\nu - x_{i_\nu}\| \leq \frac{1}{\nu}$, $\nu = 1,2,\dots$.

Hence

$$\begin{aligned}\|Ax_{i_\nu} - b\| &= \|Ax_{i_\nu} - Ax^\nu + Ax^\nu - b\| \\ &\leq \|A\| \cdot \|x_{i_\nu} - x^\nu\| + \|Ax^\nu - b\| \\ &\leq \|A\| \cdot \frac{1}{\nu} + \|Ax^\nu - b\|\end{aligned}$$

and, therefore

$$\|Ax_{i_\nu} - b\| \to \inf_{x\geq 0} \|Ax-b\| \quad \text{as} \quad \nu \to \infty,$$

which implies the desired inequality.

Now for any $x_i \in \mathscr{D}$, $\|Ax_i - b\|$ is a Borel measurable function on $\mathscr{R}$, since it is continuous. Therefore, by Th. 0.19,

$$\inf_{x_i \in \mathscr{D}} \|Ax_i - b\|$$

is also a Borel measurable function and consequently, the set

$$\mathfrak{M} = \{(A,b,c) \mid \inf_{x_i \in \mathscr{D}} \|Ax_i - b\| > 0\}$$

is Borel measurable.

b) $\mathfrak{N}$ is a Borel measurable subset of $\mathscr{R}$.
Let $B_1, \ldots, B_r$ be all $(m \times m)$-submatrices of A, i.e. $r = \binom{n}{m}$.
Then

$$\mathfrak{N} = \bigcap_{i=1}^{r} \{(A,b,c) \mid \det B_i = 0\} - \mathfrak{M}.$$

From this relation the measurability of $\mathfrak{N}$ follows immediately, since $\det B_i$ is a continuous and therefore Borel measurable function.

c) $\mathfrak{Z}$ is a Borel measurable subset of $\mathscr{R}$.
This statement is now trivial, since

$$\mathfrak{Z} = \mathscr{R} - \mathfrak{M} - \mathfrak{N}.$$

To define $\hat{x}$, we have to determine a further disjoint partition of $\mathfrak{Z}$ into measurable sets:

$$\mathfrak{C}_i = \{(A,b,c) \mid \det B_i \neq 0,\ B_i^{-1} b \geq 0\} - \bigcup_{k=1}^{i-1} \mathfrak{C}_k,\ i = 1, \ldots, r,$$

$$\mathfrak{D}_{ij} = \{(A,b,c) \mid (A,b,c) \in \mathfrak{C}_i,\ c_j - \tilde{c}'_{B_i} B_i^{-1} A_j < 0,\ B_i^{-1} A_j \leq 0\} - \bigcup_{k=1}^{j-1} \mathfrak{D}_{ik},\ j = 1, \ldots, n,\quad i = 1, \ldots, r,$$

where c_j is the j-th component of c, $\tilde{c}_{B_i}$ is the vector of components of c belonging to the basis B_i, and A_j is the j-th column of A; finally,

$$\mathfrak{E}_i = \{(A,b,c) \mid (A,b,c) \in \mathfrak{C}_i,\ c' - \tilde{c}_{B_i}{}' B_i^{-1} A \geq 0\}, \qquad i = 1, \ldots, r.$$

By definition we have a finite number of disjoint sets $\mathfrak{E}_i$ and $\mathfrak{D}_{ij}$. To show that

$$\mathfrak{Z} = \bigcup_{i=1}^{r} (\mathfrak{E}_i \cup \bigcup_{j=1}^{n} \mathfrak{D}_{ij}),$$

suppose (A,b,c) to be in $\mathfrak{Z}$. Then the linear program (1) has either a finite optimal solution and, therefore also an optimal feasible basic solution implying $(A,b,c) \in \mathfrak{E}_i$ for some i, or the objective is not bounded from below implying that there must be a feasible basic solution — i.e. $B_i^{-1} b \geq 0$ — so that some non-basic variable may be augmented arbitrarily without violating nonnegativity — i.e. $B_i^{-1} A_j \leq 0$ — and thereby decreasing the objective arbitrarily — i.e. $c_j - \tilde{c}'_{B_i} B_i^{-1} A_j < 0$. Hence $(A,b,c) \in \mathfrak{D}_{ij}$ for some (i,j). Conversely it is clear that

$$\bigcup_{i=1}^{r} (\mathfrak{E}_i \cup \bigcup_{j=1}^{n} \mathfrak{D}_{ij}) \subset \mathfrak{Z}.$$

To define $\hat{x}$ on $\mathfrak{Z}$ we shall represent this vector on $\mathfrak{E}_i \cup \bigcup_{j=1}^{n} \mathfrak{D}_{ij}$ by its basic and non-basic parts $\hat{x}^i, \hat{y}^i$ respectively. Then

$$\hat{x}^i = \begin{cases} B_i^{-1}b & \text{if } (A,b,c)\in\mathfrak{E}_i \\ B_i^{-1}b - B_i^{-1}A_j\hat{y}_j^i & \text{if } (A,b,c)\in\mathfrak{D}_{ij} \end{cases}$$

and

$$\hat{y}_\nu^i = \begin{cases} 0 \quad \text{for all } \nu, & \text{if } (A,b,c)\in\mathfrak{E}_i \\ \begin{cases} \infty & \text{for } \nu=j \\ 0 & \text{for } \nu\neq j, \end{cases} & \text{if } (A,b,c)\in\mathfrak{D}_{ij}, \end{cases}$$

which defines $\hat{x}$ on $\mathfrak{Z}$ as a Borel measurable transformation yielding a solution to (1), whenever (1) has a solution and A has full rank. Furthermore we define $\hat{x}$ on $\mathfrak{M}$ by

$$\hat{x}_i = \begin{Bmatrix} +\infty & \text{if } c_i \geq 0 \\ -\infty & \text{if } c_i < 0 \end{Bmatrix}, \quad \text{if } (A,b,c)\in\mathfrak{M}.$$

On $\mathfrak{N}$ we may define $\hat{x}$ in the same way as on $\mathfrak{Z}$ taking into account the fact that, for $(A,b,c)\in\mathfrak{N}$, a certain subset of (linearly dependent) constraints of (1) can be deleted yielding a new linear program with $m_1 < m$ linear independent constraints. □
In a certain sense Th. 2 is stronger than Th. 1.

Precisely:

Theorem 3. *Let $\hat{x}$ be the measurable transformation of Th. 2 and $\hat{\gamma}=c'\hat{x}$. Then $\hat{\gamma}$ is a Borel measurable extended real valued function on $\mathbb{R}^{m\times n+m+n}$ and $\hat{\gamma}=\gamma$ almost everywhere with respect to the Lebesgue measure.*

Proof. Every component $\hat{x}_i$ of $\hat{x}$ is a Borel measurable function on $\mathbb{R}^{m\times n+m+n}$ as well as every component c_i of c. Therefore $\hat{\gamma}=c'\hat{x}$ as a sum of products of measurable functions is measurable (see Th. 0.19).
$\hat{\gamma}$ may differ from γ just on the set

$$\mathscr{G} = \{(A,b,c) \mid (A,b,c)\in\mathfrak{M},\ c=0\},$$

because there $\hat{\gamma}=0$ (if we use the definition $0\cdot\infty=0$) and $\gamma=\infty$.
But the Lebesgue measure of $\mathscr{G}$ is equal to zero. □

Although these results indicate that, from a mathematical point of view, the distribution problem is meaningful, we may still get distribution functions of the optimal value with defect, i.e. it may happen that the probability

$$P_\omega(\{\omega \mid -\infty < \gamma < \infty\}) < 1.$$

In this case the moments of the random variable γ certainly do not exist. Since, however, in many practical cases decision makers are interested in the mean value and the variance, we shall try to characterize those problems whose optimal value has a distribution function without defect. After that we shall investigate some special types of problems which are of practical importance as well as mathematically "well behaved". The following results are due to Bereanu [4].

Theorem 4. $P_\omega(\{\omega \mid -\infty<\gamma<\infty\})=1$ *if and only if the implications:*

a) *for all u, if* $u'A \geq 0$ *then* $b'u \geq 0$

b) *for all* $w \geq 0$, *if* $Aw=0$ *then* $c'w \geq 0$

hold with probability 1.

Proof. $P_\omega(\{\omega \mid -\infty<\gamma<\infty\})=1$ if and only if

a) (1) has feasible solutions with probability 1,

b) the objective $c'x$ of (1) is unbounded from below on the feasible set with probability 0.

If we define

$\mathfrak{B}=\{(A,b,c) \mid Ax=b,\ x \geq 0$ has feasible solutions$\}$

$=\mathscr{R}-\mathfrak{M}$ in the terminology of Th. 2

and

$\mathfrak{U}=\{(A,b,c) \mid Ax=b,\ x \geq 0,$ has feasible solutions such that $c'x$ is unbounded from below$\}$

$=\{(A,b,c) \mid (A,b,c) \in \mathfrak{B}$ and there exist $w \geq 0$ such that $Aw=0$ and $c'w<0\}$,

then we must have

$$\text{a) } P_\omega(\mathfrak{B})=1$$
$$\text{b) } P_\omega(\mathfrak{U})=0.$$

By Farkas' lemma (see Th. 0.1)

$(A,b,c) \in \mathfrak{B}$ if and only if $u'A \geq 0$ implies $b'u \geq 0\ (u \in \mathbb{R}^m)$,

and therefore

$$P_\omega(\mathfrak{B})=P_\omega(\{(A,b,c) \mid u'A \geq 0 \quad \text{implies} \quad b'u \geq 0\})=1.$$

Finally, we need $P_\omega(\mathfrak{U})=0$ or equivalently $P_\omega(\mathscr{R}-\mathfrak{U})=1$.

Since

$$\mathfrak{U}=\mathfrak{B} \cap \{(A,b,c) \mid \exists w \geq 0, Aw=0 \quad \text{implies} \quad c'w<0\},$$

we get

$$\mathscr{R}-\mathfrak{U}=(\mathscr{R}-\mathfrak{B}) \cup \{(A,b,c) \mid w \geq 0, Aw=0 \quad \text{implies} \quad c'w \geq 0\},$$

and hence

$$1=P_\omega(\mathscr{R}-\mathfrak{U}) \leq P_\omega(\mathscr{R}-\mathfrak{B})+P_\omega(\{(A,b,c) \mid w \geq 0, Aw=0 \quad \text{implies} \quad c'w \geq 0\}).$$

The fact, that $P_\omega(\mathfrak{B})=1$ and consequently $P_\omega(\mathscr{R}-\mathfrak{B})=0$, completes the proof. □

Let us now consider two special types of linear programs, which satisfy the conditions of Th. 4 and hence yield an optimal value, whose distribution function has no defect.

$$\text{(3)} \qquad \begin{aligned} \gamma &= \inf c'x \\ Ax &\geq b \\ x &\geq 0 \end{aligned}$$

where

$$\sum_{j=1}^{n} a_{ij}>0, \qquad i=1,\ldots,m,$$

$$\sum_{i=1}^{m} a_{ij}>0, \qquad j=1,\ldots,n \quad \text{and}$$

$$c_j \geq 0, \qquad j=1,\ldots,n, \quad \text{with probability 1.}$$

We call this type of linear program a *positive linear program*. The assumptions made here are quite realistic if we imagine that (3) represents a production program, where A is the technological matrix, x represents the input, Ax the output, b the demand vector and c_j the cost per unit of the j-th factor of production.

Theorem 5. *For any positive linear program* (3),

$$P_\omega(\{\omega \mid -\infty < \gamma < \infty\}) = 1.$$

Proof. Introducing slack variables we get

$$\begin{aligned} \gamma &= \inf d'z \\ Bz &= b \\ z &\geq 0 \end{aligned}$$

where

$$\begin{aligned} d' &= (c', 0') \\ B &= (A, -E) \\ z' &= (x', y') \end{aligned}$$

To verify condition a) of Th. 4, let $u \in \mathbb{R}^m$ be such that $u'B \geq 0$, i.e. $u'A \geq 0$ and $u' \leq 0$.
Since (3) is positive, it follows that $u = 0$ and therefore that $b'u = 0$. Condition b) of Th. 4 follows from the fact that $w \geq 0$ already implies $d'w \geq 0$. □

Another special type is the *stochastic transportation problem*, where the unit transportation costs, the supplies and the demands are assumed to be random variables with positive range and such that the total demand almost surely does not exceed the total supply.

$$\gamma = \inf \sum_{i=1}^{m} \sum_{j=1}^{n} c_{ij} x_{ij}$$

(4)
$$\sum_{j=1}^{n} x_{ij} \leq a_i, \quad i = 1, \ldots, m$$

$$\sum_{i=1}^{m} x_{ij} \geq b_j, \quad j = 1, \ldots, n$$

$$x_{ij} \geq 0,$$

where $c_{ij} \geq 0$, $a_i \geq 0$, $b_j \geq 0$ and $\sum_{j=1}^{n} b_j \leq \sum_{i=1}^{m} a_i$ with probability 1.

Theorem 6. *The distribution function $F_\gamma(\xi)$ of the stochastic transportation problem* (4) *has no defect, i.e.*

$$P_\omega(\{\omega \mid -\infty < \gamma < \infty\}) = 1.$$

Proof. Introducing slack variables, we get the matrix

$$A=\left(\begin{array}{cccc|c|c} d' & & & & & \\ & d' & & & & \\ & & \ddots & & I_m & \\ & & & d' & & \\ \hline I_n & I_n & \cdots & I_n & & -I_n \end{array}\right)$$

where $d'=\underbrace{(1,1,\ldots,1)}_{n\ \text{times}}$ and I_r is the $(r\times r)$-identity matrix.

Let $u\in\mathbb{R}^{m+n}$, i.e.

$$u=\begin{pmatrix}u^1\\u^2\end{pmatrix}\quad\text{with}\quad u^1\in\mathbb{R}^m,\qquad u^2\in\mathbb{R}^n.$$

From $u'A\geq 0$ it follows that $u^1\geq 0$, $u^2\leq 0$ and $u_i^1+u_j^2\geq 0$, for $i=1,\ldots,m, j=1,\ldots,n$, and hence, with probability 1,

$$\sum_{i=1}^m a_i u_i^1+\sum_{j=1}^n b_j u_j^2\geq \min_{1\leq i\leq m} u_i^1\cdot\sum_{i=1}^m a_i+\min_{1\leq j\leq n} u_j^2\cdot\sum_{j=1}^n b_j$$

$$\geq\left(\min_{1\leq i\leq m} u_i^1+\min_{1\leq j\leq n} u_j^2\right)\cdot\sum_{j=1}^n b_j\geq 0,$$

which coincides with condition a) of Th. 4.
On the other hand, condition b) of Th. 4 is trivially satisfied, since for $w'=(w^{1'},w^{2'})\geq 0$, where $w^1\in\mathbb{R}^{m\cdot n}$ and $w^2\in\mathbb{R}^{m+n}$, $c'w^1\geq 0$. □

It often happens, as in problems (3) and (4), that only a certain subset of the coefficients of a linear program are random variables. We may express this fact by a reformulation of the general problem due to Bereanu [3], which allows statements of all possible kinds of SLP distribution problems.
Let $T\subset\mathbb{R}^r$ be the range of a random vector $t=(t_1,\ldots,t_r)'$ and

$$\text{(5)}\qquad \begin{aligned} \gamma(t)&=\inf c'(t)\cdot x\\ \text{subject to } A(t)x&=b(t)\\ x&\geq 0 \end{aligned}$$

where

$$\begin{aligned} A(t)&=A^0+A^1\cdot t_1+A^2\cdot t_2+\cdots+A^r t_r\\ b(t)&=b^0+b^1\cdot t_1+b^2\cdot t_2+\cdots+b^r\cdot t_r\\ c(t)&=c^0+c^1\cdot t_1+c^2\cdot t_2+\cdots+c^r\cdot t_r \end{aligned}$$

with deterministic real matrices A^i, b^i, c^i.
By Th. 1 it is obvious, that $\gamma(t)$, if it exists, is Borel measurable in t, since $A(t)$, $b(t)$, $c(t)$ are continuous in t.

Theorem 7. *Let $B(t)$ be an $(m \times m)$-submatrix of $A(t)$ and μ_r the Lebesgue measure on $\mathbb{R}^r$. Then either* $\det B(t)=0$ *for all $t \in T$ or* $\mu_r(\{t \mid \det B(t)=0\})=0$.

Proof. Suppose it is not the case that $\det B(t)=0$ for all $t \in T$, i.e. there exists a $t^* \in T$ such that $\det B(t^*) \neq 0$.
Obviously $\varphi(t)=\det B(t)$ is an algebraic function in t, i.e.

$$\varphi(t)=\sum_{\nu=1}^{\varkappa} \alpha_\nu t_1^{i_{1\nu}} t_2^{i_{2\nu}} \ldots t_r^{i_{r\nu}}$$

where $i_{r\nu} \geq 0$ are integers. Now for any algebraic function $\varphi(t)$ either $\varphi(t) \equiv 0$ or $\mu_r(\{t \mid \varphi(t)=0\})=0$, as we shall see by induction to r.
For $r=1$ the fundamental theorem of algebra asserts that $\varphi(t) \equiv 0$ or $\varphi(t)$ has a finite number of roots, i.e. $\mu_1(\{t \mid \varphi(t)=0\})=0$.
Assume that the statement is true for any algebraic function of at most $r-1$ variables $t_1, \ldots, t_{r-1}$. Consider now

$$\varphi(t)=\sum_{\nu=1}^{\varkappa} \alpha_\nu t_1^{i_{1\nu}} t_2^{i_{2\nu}} \ldots, t_r^{i_{r\nu}}.$$

Then either $\varphi(t) \equiv 0$ or

$$\varphi(t)=\sum_{\mu=0}^{A} \varphi_\mu(t_1, \ldots, t_{r-1}) \cdot t_r^\mu,$$

where $\varphi_\mu(t_1, \ldots, t_{r-1})$ are algebraic functions of at most $r-1$ variables $t_1, \ldots, t_{r-1}$ such that there exists at least one μ_0 with $\varphi_{\mu_0}(t_1, \ldots, t_{r-1}) \not\equiv 0$.

$$\begin{aligned} \{t \mid \varphi(t)=0\} = {} & [\{t \mid \varphi_{\mu_0}(t_1, \ldots, t_{r-1})=0\} \cap \{t \mid \varphi(t)=0\}] \cup \\ & \cup [\{t \mid \varphi_{\mu_0}(t_1, \ldots, t_{r-1}) \neq 0\} \cap \{t \mid \varphi(t)=0\}] \\ = {} & \mathfrak{C} \cup \mathfrak{D} \end{aligned}$$

Now for any t_r-section $\mathfrak{C}_{t_r}$ of $\mathfrak{C}$ we have, by assumption,

$$\mu_{r-1}(\mathfrak{C}_{t_r}) \leq \mu_{r-1}(\{(t_1, \ldots t_{r-1}) \mid \varphi_{\mu_0}(t_1, \ldots, t_{r-1})=0\})=0$$

which implies $\mu_r(\mathfrak{C})=0$ by Th. 0.25. And for any $(t_1, \ldots, t_{r-1})$-section $\mathfrak{D}_t^{(r-1)}$ of $\mathfrak{D}$ we get

$$\begin{aligned} \mu_1(\mathfrak{D}_t^{(r-1)}) & = \mu_1\left(\{t_r \mid \sum_{\mu=0}^{A} \varphi_\mu(t_1, \ldots, t_{r-1}) \cdot t_r^\mu = 0;\ \varphi_{\mu_0}(t_1, \ldots, t_{r-1}) \neq 0\}\right) \\ & = 0, \end{aligned}$$

again by the fundamental theorem of algebra, implying $\mu_r(\mathfrak{D})=0$ by Th. 0.25. □

Motivated by this theorem, we call any submatrix $B(t)$ of $A(t)$ *almost nonsingular*, if it is nonsingular for at least one $t \in T$. To give a general formula for the distribution function of $\gamma(t)$, let us assume, that

A1 a) the probability measure P_t on T is absolutely continuous with respect to the Lebesgue measure μ_r, i.e. there exists a probability density function $f_t(\tau)$;

A1 b) in (5), $A(t)$, $b(t)$ and $c(t)$ satisfy the conditions a) and b) of Th. 4;

A1 c) there exists a $t^* \in T$ such that $A(t^*)$ has rank m.

Let $\{B_i(t) | i=1,\ldots,q\}$ be the class of all almost nonsingular $(m \times m)$-submatrices of $A(t)$. By assumption A1 c) this class is not empty.
For simplicity let us define arbitrarily $B_i^{-1}(t)=0$ for all $t \in T$ with $\det B_i(t)=0$, $i=1,\ldots,q$.
By Th. 7, this definition is valid on a t-set of Lebesgue measure zero and hence, by assumption A1 a), of probability zero. Furthermore we define

$$\mathfrak{A}_i=\{t \mid t \in T; B_i^{-1}(t)b(t) \geq 0 \quad \text{and} \quad c'(t)-c'_{B_i}(t)B_i^{-1}(t)A(t) \geq 0\}, \qquad i=1,\ldots,q,$$

where $c_{B_i}(t)$ consists of the basic components of $c(t)$ with respect to the basis $B_i(t)$. Moreover let

$$\mathfrak{B}_i=\mathfrak{A}_i-\bigcup_{j=1}^{i-1}\mathfrak{A}_j; \quad i=1,\ldots,q; \text{ i.e. } \mathfrak{B}_i \cap \mathfrak{B}_k=\emptyset, \quad i \neq k, \text{ and } \bigcup_{i=1}^{q}\mathfrak{B}_i=\bigcup_{i=1}^{q}\mathfrak{A}_i.$$

It is now easy to prove

Theorem 8. *Assume A1. Then* $P_t\left(\bigcup_{i=1}^{q}\mathfrak{B}_i\right)=\sum_{i=1}^{q}P_t(\mathfrak{B}_i)=1.$

Let $\gamma_i(t):\mathfrak{B}_i \to \mathbb{R}$ *be defined as* $\gamma_i(t)=c'_{B_i}(t)B_i^{-1}(t)b(t)$ *and* $\mathfrak{C}_i(\xi)=\{t \mid t \in \mathfrak{B}_i$ *and* $\gamma_i(t) \leq \xi\}$. *Then the distribution* $F_{\gamma(t)}(\xi)$ *of* $\gamma(t)$ *is determined by*

$$F_{\gamma(t)}(\xi)=\sum_{i=1}^{q}\int_{\mathfrak{C}_i(\xi)}f_t(\tau)\,d\tau.$$

Proof. Let

$$\mathfrak{D}=\{t \mid t \in T; \ -\infty<\gamma(t)<\infty\}$$
$$\mathfrak{E}=\{t \mid t \in T; \ \text{rank } (A(t))=m\}; \text{ then}$$
$$P_t(\mathfrak{D})=1 \text{ and } P_t(\mathfrak{E})=1 \text{ by A1 b) and c) respectively and Th. 7.}$$

For $t \in \mathfrak{D} \cap \mathfrak{E}$ there exists a basis $B_i(t)$ so that $t \in \mathfrak{A}_i$; hence $\mathfrak{D} \cap \mathfrak{E} \subset \bigcup_{i=1}^{q}\mathfrak{A}_i=\bigcup_{i=1}^{q}\mathfrak{B}_i$.

Therefore

$$\begin{aligned}1 \geq P_t\left(\bigcup_{i=1}^{q}\mathfrak{B}_i\right) \geq P_t(\mathfrak{D} \cap \mathfrak{E}) &= P_t(\mathfrak{D}-(\mathfrak{D}-\mathfrak{E}))\\ &= P_t(\mathfrak{D})-P_t(\mathfrak{D}-\mathfrak{E})\\ &\geq P_t(\mathfrak{D})-P_t(T-\mathfrak{E})\\ &= P_t(\mathfrak{D})-P_t(T)+P_t(\mathfrak{E})=1\end{aligned}$$

By construction from $\mathfrak{B}_i \cap \mathfrak{B}_k=\emptyset$, $i \neq k$, it follows that

$$P_t\left(\bigcup_{i=1}^{q}\mathfrak{B}_i\right)=\sum_{i=1}^{q}P_t(\mathfrak{B}_i)=1.$$

We are interested in the distribution function $F_{\gamma(t)}(\xi)$ of $\gamma(t)$, i.e. in the probability of the set

$$\mathscr{G}(\xi)=\{t \mid t \in T; \ -\infty<\gamma(t) \leq \xi\}; \qquad \xi \in \mathbb{R}.$$

Let $\mathfrak{B}=\bigcup_{i=1}^{q}\mathfrak{B}_i$. Then obviously $P_t(\mathscr{G}(\xi)) \geq P_t(\mathscr{G}(\xi) \cap \mathfrak{B})$.

On the other hand

$$\begin{aligned} P_t(\mathcal{G}(\xi)\cap\mathfrak{B}) &= P_t(\mathcal{G}(\xi) - (\mathcal{G}(\xi)-\mathfrak{B})) \\ &= P_t(\mathcal{G}(\xi)) - P_t(\mathcal{G}(\xi)-\mathfrak{B}) \\ &\geq P_t(\mathcal{G}(\xi)) - P_t(T-\mathfrak{B}) \\ &= P_t(\mathcal{G}(\xi)), \quad \text{since} \quad P_t(\mathfrak{B})=1, \end{aligned}$$

and hence

$$P_t(\mathcal{G}(\xi)) = P_t(\mathcal{G}(\xi)\cap\mathfrak{B}) = \sum_{i=1}^{q} P_t(\mathcal{G}(\xi)\cap\mathfrak{B}_i).$$

Let $\mathfrak{B}_i^{(0)} = \{t \mid \det B_i(t)=0 \text{ and } t\in\mathfrak{B}_i\}, \quad i=1,\ldots,q.$
Since $B_i(t)$ is almost nonsingular, we have

$$\mu_r(\mathfrak{B}_i^{(0)})=0, \qquad i=1,\ldots,q,$$

and in cause of the absolute continuity of P_t to μ_r,

$$P_r(\mathfrak{B}_i^{(0)})=0 \quad \text{and}$$
$$P_t(\mathcal{G}(\xi)\cap\mathfrak{B}_i) = P_t(\mathcal{G}(\xi)\cap(\mathfrak{B}_i-\mathfrak{B}_i^{(0)})), \qquad i=1,\ldots,q.$$

But, for $t\in\mathfrak{B}_i-\mathfrak{B}_i^{(0)}$, we have $B_i(t)$ as an optimal feasible basis, i.e.

$$\gamma(t) = c'_{B_i(t)}B_i^{-1}(t)\cdot b(t) = \gamma_i(t).$$

Therefore

$$\begin{aligned} F_{\gamma(t)}(\xi) = P_t(\mathcal{G}(\xi)) &= \sum_{i=1}^{q} P_t(\mathcal{G}(\xi)\cap(\mathfrak{B}_i-\mathfrak{B}_i^{(0)})) \\ &= \sum_{i=1}^{q} P_t(\{t \mid t\in\mathfrak{B}_i-\mathfrak{B}_i^{(0)} \quad \text{and} \quad \gamma_i(t)\leq\xi\}) \\ &= \sum_{i=1}^{q} P_t(\{t \mid t\in\mathfrak{B}_i \quad \text{and} \quad \gamma_i(t)\leq\xi\}), \quad \text{since} \quad P_t(\mathfrak{B}_i^{(0)})=0, \\ &= \sum_{i=1}^{q} P_t(\mathfrak{C}_i(\xi)) = \sum_{i=1}^{q} \int_{\mathfrak{C}_i(\xi)} f_t(\tau)\,d\tau. \square \end{aligned}$$

The following proposition is an immediate consequence of Th. 8:

Theorem 9. *With the assumptions and notations of Th. 8 the ν-th moment of $\gamma(t)$ is determined, provided that the integrals exist, by*

$$E\gamma^{\nu}(t) = \sum_{i=1}^{q} \int_{\mathfrak{B}_i} \gamma_i^{\nu}(\tau) f_t(\tau)\,d\tau.$$

Proof. As we have seen in Th. 8,

$$P_t\left(\bigcup_{i=1}^{q}\mathfrak{B}_i\right) = \sum_{i=1}^{q} P_t(\mathfrak{B}_i)=1,$$

$$\gamma(t)=\gamma_i(t) \quad \text{for} \quad t\in\mathfrak{B}_i-\mathfrak{B}_i^{(0)} \quad \text{and} \quad \mu_r(\mathfrak{B}_i^{(0)})=P_t(\mathfrak{B}_i^{(0)})=0.$$

Hence,

$\gamma(t)=\gamma_i(t)$ almost everywhere on $\mathfrak{B}_i$ with respect to μ_r and P_t.

This completes the proof. □

Sometimes one can find the conjecture in the literature, that under the assumptions A1, or even stronger ones like positivity assumptions on the linear program (5), the sets $\mathfrak{A}_i$ could be taken as "decision regions" instead of the sets $\mathfrak{B}_i$. The following very simple and well-behaved example shows, that this conjecture cannot be true in general because

$$P_t(\mathfrak{A}_i \cap \mathfrak{A}_k) = 0, \qquad i \neq k,$$

is not true in general, in spite of such assumptions.
Consider

$$\gamma(t) = \inf\left(2 + \frac{1}{2}t\right)x_1 + 3x_2$$
$$\text{subject to } \left(5 + \frac{5}{4}t\right)x_1 + \frac{15}{2}x_2 = 5 + \frac{1}{3}t$$
$$x_1 \geq 0, \qquad x_2 \geq 0$$

and let t have a probability density function on

$$T = \{t \mid 0 \leq t \leq 4\} \subset \mathbb{R}.$$

Then

$$B_1 = \left(5 + \frac{5}{4}t\right), \qquad B_2 = \left(\frac{15}{2}\right)$$

$$\mathfrak{A}_1 = \left\{t \mid t \in [0,4], \frac{5 + \frac{1}{3}t}{5 + \frac{5}{4}t} \geq 0 \text{ and } 3 - \frac{\left(2 + \frac{1}{2}t\right) 15}{\left(5 + \frac{5}{4}t\right) \cdot 2} \geq 0\right\} = T$$

and

$$\mathfrak{A}_2 = \left\{t \mid T \in [0,4], \frac{\left(5 + \frac{1}{3}t\right) \cdot 2}{15} \geq 0 \text{ and } \left(2 + \frac{1}{2}t\right) - \frac{3 \cdot 2\left(5 + \frac{5}{4}t\right)}{15} \geq 0\right\} = T$$

and hence,

$$P_t(\mathfrak{A}_1 \cap \mathfrak{A}_2) = P_t(T) = 1.$$

Therefore one should be very careful in replacing $\mathfrak{B}_i$ by $\mathfrak{A}_i$ in Th. 8 and 9. However, $P_t(\mathfrak{A}_i \cap \mathfrak{A}_k) = 0$, $i \neq k$, is a sufficient condition for replacing $\mathfrak{B}_i$ by $\mathfrak{A}_i$ to be allowed.

Theorem 10. *Assume A1. Then* $\mu_r(\mathfrak{A}_i \cap \mathfrak{A}_k) = P_t(\mathfrak{A}_i \cap \mathfrak{A}_k) = 0$, $i \neq k$, $i, k \in \{1, \ldots, q\}$, *if either*

$$\mu_r(\{t \mid t \in T, c'_{B_i}(t) B_i^{-1}(t) b(t) - c'_{B_k}(t) B_k^{-1}(t) b(t) = 0\}) = 0$$
$$\textit{or} \quad \mu_r(\{t \mid t \in T, B_i^{-1}(t) b(t) \geq 0 \quad \textit{and} \quad B_k^{-1}(t) b(t) \geq 0\}) = 0.$$

Proof. By definition

$$\mathfrak{A}_i = \{t \mid t \in T, B_i^{-1}(t) b(t) \geq 0, c'(t) - c'_{B_i}(t) B_i^{-1}(t) A(t) \geq 0\},$$

and hence $\quad t \in \mathfrak{A}_i \cap \mathfrak{A}_k \quad$ if only if

$$t \in T \cap \{t \mid B_i^{-1}(t)b(t) \geq 0,\ B_k^{-1}(t)b(t) \geq 0\} \cap$$
$$\cap \{t \mid c'(t) - c'_{B_i}(t)B_i^{-1}(t)A(t) \geq 0;\ c'(t) - c'_{B_k}(t)B_k^{-1}(t)A(t) \geq 0\}.$$

Therefore, for $t \in \mathfrak{A}_i \cap \mathfrak{A}_k$ we have in particular,

$$c'_{B_k}(t) - c'_{B_i}(t)B_i^{-1}(t)B_k(t) \geq 0 \quad \text{and} \quad B_k^{-1}(t)b(t) \geq 0$$

and

$$c'_{B_i}(t) - c'_{B_k}(t)B_k^{-1}(t)B_i(t) \geq 0 \quad \text{and} \quad B_i^{-1}(t)b(t) \geq 0,$$

which implies

$$c'_{B_i}(t)B_i^{-1}(t)b(t) - c'_{B_k}(t)B_k^{-1}(t)b(t) = 0.$$

Hence,

$$\mathfrak{A}_i \cap \mathfrak{A}_k \subset \{t \mid t \in T,\ c'_{B_i}(t)B_i^{-1}(t)b(t) - c'_{B_k}(t)B_k^{-1}(t)b(t) = 0\}$$

and obviously,

$$\mathfrak{A}_i \cap \mathfrak{A}_k \subset \{t \mid t \in T,\ B_i^{-1}(t)b(t) \geq 0 \quad \text{and} \quad B_k^{-1}(t)b(t) \geq 0\},$$

which proves the theorem. □

From the counter example above as well as from the last proof we might suggest that there is some connection between the fact that $t \in \mathfrak{A}_i \cap \mathfrak{A}_k$ and primal or dual degeneracy. Let $B_i(t)$ again be some almost nonsingular $(m \times m)$-submatrix of $A(t)$. We say that $B_i(t)$ is primal degenerated with respect to the linear program (5), if at least one component of $B_i^{-1}(t)b(t)$ vanishes. We call $B_i(t)$ dual degenerated, if at least one component of $c_{R_i}(t) - R_i'(t)B'^{-1}(t)c_{B_i}(t)$ is equal to zero, where $R_i(t)$ is the matrix consisting of all columns of $A(t)$ not belonging to $B_i(t)$ and $c_{R_i}(t)$ is the vector of those components of $c(t)$ belonging to $R_i(t)$. Certain stochastic linear programs of type (5) satisfy the assumption

A2. For every almost nonsingular $(m \times m)$-submatrix $B_i(t)$ of $A(t)$ there exists a $t^{(i)} \in T$ such that $B_i(t^{(i)})$ is nonsingular and neither primal nor dual degenerated with respect to (5).

Theorem 11. *Given A1 and A2, we have* $P_t(\mathfrak{A}_i \cap \mathfrak{A}_k) = 0$, $i \neq k$, i.e. *the sets* $\mathfrak{A}_i$ *may be taken as "decision regions".*

Proof.
$$\begin{aligned}\mathfrak{A}_i \cap \mathfrak{A}_k = \{t \mid & B_i^{-1}(t)b(t) \geq 0,\ B_k^{-1}(t)b(t) \geq 0,\\ & c'(t) - c'_{B_i}(t)B_i^{-1}(t)A(t) \geq 0,\\ & c'(t) - c'_{B_k}(t)B_k^{-1}(t)A(t) \geq 0,\ t \in T\}\\ \subset \{t \mid & B_i^{-1}(t)b(t) \geq 0,\ B_k^{-1}(t)b(t) \geq 0,\\ & c'_{B_i}(t) - c'_{B_k}(t)B_k^{-1}(t)B_i(t) \geq 0,\\ & c'_{B_i}(t)B_i^{-1}(t)b(t) - c'_{B_k}(t)B_k^{-1}(t)b(t) = 0,\ t \in T\} = \mathfrak{D}\end{aligned}$$

as we know from the proof of Th. 10. Obviously

$$\mathfrak{D} = [\{t \mid \det B_k(t) = 0\} \cap \mathfrak{D}] \cup [\{t \mid \det B_k(t) \neq 0\} \cap \mathfrak{D}]$$

and

$$\mu_r(\mathfrak{D}) = \mu_r(\mathfrak{D} \cap \{t \mid \det B_k(t) \neq 0\})$$

since $B_k(t)$ is almost nonsingular and, for the same reason,

$$\mu_r(\mathfrak{D})=\mu_r(\mathfrak{D}\cap\{t\mid\det B_k(t)\neq 0\}\cap\{t\mid\det B_i(t)\neq 0\}).$$

For det $B_i(t)\neq 0$ define

$$D_i(t)=\det B_i(t)\cdot B_i^{-1}(t),$$

i.e. the elements of the matrix $D_i(t)$ are algebraic functions in t.

For
$$t\in\mathfrak{D}\cap\{t\mid\det B_k(t)\neq 0\}\cap\{t\mid\det B_i(t)\neq 0\}=\mathfrak{C}$$
$$c'_{B_i}(t)B_i^{-1}(t)b(t)-c'_{B_k}(t)B_k^{-1}(t)b(t)=0$$

if and only if at least one of any two corresponding components of the vectors

$$c'_{B_i}(t)-c'_{B_k}(t)B_k^{-1}(t)B_i(t) \quad \text{and}$$
$$B_i^{-1}(t)b(t)=\frac{1}{\det B_i(t)}\cdot D_i(t)b(t)$$

vanishes. By assumption A2 any component of $D_i(t)b(t)$ is an algebraic function in t, not vanishing in $t^{(i)}$ and therefore vanishing only on a t-set of Lebesgue measure zero.

In $\mathfrak{C}$, a component of

$$c'_{B_i}(t)-c'_{B_k}B_k^{-1}(t)B_i(t)$$

vanishes if and only if the same component of

$$d(t)=\det B_k(t)\cdot c'_{B_i}(t)-c'_{B_k}D_k(t)B_i(t)$$

vanishes. Since $B_i(t)$ and $B_k(t)$ are different submatrices of $A(t)$, there is at least one component of $d(t)$, say $d_\nu(t)$, which is an algebraic function in t and, by A2, $d_\nu(t^{(k)})\neq 0$ and therefore

$$\mu_r(\{t\mid d_\nu(t)=0\})=0.$$

This yields the desired result

$$\mu_r(\mathfrak{D})=0$$

and hence

$$P_t(\mathfrak{A}_i\cap\mathfrak{A}_k)=0. \square$$

2. Special Problems

From Th. 8 and Th. 9 we may conclude that in general it is not at all trivial to determine the distribution function $F_{\gamma(t)}(\xi)$ or the moments $E\gamma^\nu(t)$, since its computation involves numerical integration over the sets $\mathfrak{C}_i(\xi)$ or functions $\gamma_i^\nu(t)$ which are difficult to handle. One of the major reasons for these difficulties is the fact that, in general, $\gamma(t)$ is not continuous in t.

The following example due to Bereanu [4] shows that this discontinuity may appear even if $\gamma(t)$ is finite for all $t \in T$ as soon as the technological matrix varies with t.
Define, for $t \in \mathbb{R}$,

$$\gamma(t) = \inf\{x \mid x \in \mathbb{R},\quad y \in \mathbb{R},\quad x + ty \geq 1,\quad x \geq 0,\quad y \geq 0\}.$$

Then

$$\gamma(t) = \begin{cases} 1 & \text{for} \quad t \leq 0 \\ 0 & \text{for} \quad t > 0, \end{cases}$$

hence, $\gamma(t)$ exists everywhere in t, but it is not continuous at $t=0$.
However, there are special situations where the continuity of $\gamma(t)$ can be established. Consider the parametric program

$$\begin{aligned} &\gamma(t) = \inf c'(t)x \\ (6) \qquad &\text{subject to} \quad Ax = b(t) \\ &\qquad\qquad\qquad x \geq 0 \\ \text{where, as in (5),}\quad &c(t) = c^0 + c^1 \cdot t_1 + c^2 \cdot t_2 + \cdots + c^r \cdot t_r \\ &b(t) = b^0 + b^1 \cdot t_1 + b^2 t_2 + \cdots + b^r t_r \end{aligned}$$

but A is constant. Then we may prove

Theorem 12. *Let $T \subset \mathbb{R}^r$ be a closed interval. Assume that*

$$\{x \mid Ax = b(t),\ x \geq 0\} \neq \emptyset \quad \textit{and}$$
$$\{y \mid y \geq 0,\quad Ay = 0,\ c'(t)y < 0\} = \emptyset \quad \textit{for all } t \in T.$$

Then $\gamma(t)$ is continuous on T.

Proof. Without loss of generality we may assume that A has rank m. If $\{B_i \mid i = 1, \ldots, q\}$ is the class of all nonsingular $(m \times m)$-submatrices of A, then $\mathfrak{A}_i = \{t \mid t \in T,\ B_i^{-1} b(t) \geq 0,\ c'(t) - c'_{B_i}(t) B_i^{-1} A \geq 0\}$, $i = 1, \ldots, q$, are closed convex polyhedral sets, which according to our assumptions cover T, i.e. $T = \bigcup_{i=1}^{q} \mathfrak{A}_i$.
For $t \in \mathfrak{A}_i$ we have

$$\gamma(t) = \gamma_i(t) = c'_{B_i}(t) B_i^{-1} b(t),$$

i.e. $\gamma(t) = \gamma_i(t)$ is an algebraic function in $(t_1, \ldots, t_r)$ and therefore continuous on $\mathfrak{A}_i$. The continuity of $\gamma(t)$ on T now follows from the fact that $\gamma_i(t) = \gamma_j(t)$ for $t \in \mathfrak{A}_i \cap \mathfrak{A}_j$, since $\gamma(t)$ is uniquely determined. $\square$

If in addition to the assumptions of Th. 12 $c = c(t) \equiv c^0$, i.e. $c^1 = c^2 = \cdots = c^r = 0$, we see from this proof that $\gamma(t)$ is piecewise linear (see Th. 0.7). If moreover

$T = \{t \mid \alpha \leq t \leq \beta,\quad t \in \mathbb{R}\}$ we may proceed as follows:

(i) For $t_0 = \alpha$ determine $\gamma(t_0) = \min\{c'x \mid Ax = b(t_0),\ x \geq 0\}$ with the simplex method yielding an optimal feasible basis B_0 such that $\gamma(t_0) = c'_{B_0} B_0^{-1} b(t_0)$. With $k = 0$ go to the next step.

(ii) Define $\tau=\sup\{t \mid B_k^{-1} b(t) \geq 0\}$.
If $\tau \geq \beta$, set $t_K = t_{k+1} = \beta$ and stop.
If $\tau < \beta$, set $t_{k+1} = \tau$ and determine by the dual simplex method a new optimal feasible basis B_{k+1} so that
$B_{k+1}^{-1} b(t) \geq 0$ for $t \in \{t \mid t_{k+1} \leq t \leq t_{k+1} + \varepsilon\}$ for some $\varepsilon > 0$.
Then replace k by $k+1$ and repeat step (ii).

From our assumptions it is evident that this procedure terminates after a finite number of steps. Then we have, for $t_k \leq t \leq t_{k+1}$ and $k=0,1 \ldots K-1$,

$$\begin{aligned} \gamma(t) &= c'_{B_k} B_k^{-1} b(t) \\ &= \frac{t_{k+1}-t}{t_{k+1}-t_k} \gamma(t_k) + \frac{t-t_k}{t_{k+1}-t_k} \gamma(t_{k+1}) \end{aligned}$$

Given a probability distribution on T for example by a continuous density function $f_t(\tau)$, we may now — according to Th. 8 and Th. 9 — determine $F_{\gamma(t)}(\xi)$ or $E\gamma^\nu(t)$ by numerical quadrature. Now, it is clear that we have to define the decision regions in this case as

$$\begin{aligned} &\mathfrak{B}_k = \{t \mid t_{k-1} \leq t < t_k\} \quad \text{for} \quad k=1,\ldots,K-1 \quad \text{and} \\ &\mathfrak{B}_K = \{t \mid t_{K-1} \leq t \leq t_K\}. \end{aligned}$$

Now let us consider the general problem

$$\begin{aligned} \gamma(t) = \inf c'(t)x& \\ \text{subject to } A(t)x &= b(t) \\ x &\geq 0 \end{aligned} \tag{7}$$

as stated in (5), and assume that t varies over a compact interval $T \subset \mathbb{R}^r$. For $\gamma(t)$ to be continuous on T it is necessary that (7) is solvable for all $t \in T$, or equivalently (see Th. 0.1 and Th. 0.4) the conditions

$$\begin{aligned} &w \in \mathbb{R}^n,\ w \geq 0,\ A(t)w=0 \text{ implies } c'(t)w \geq 0 \text{ for all } t \in T \\ &u \in \mathbb{R}^m,\ A'(t)u \leq 0 \text{ implies } b'(t)u \leq 0 \text{ for all } t \in T \end{aligned} \tag{8}$$

are necessary. However, these conditions are not sufficient for the continuity of $\gamma(t)$ on T, as we know from the example given above.

To motivate a sufficient condition given by Bereanu [4], let us assume that for $(m \times m)$-submatrices $B_i(t)$, $i=1,\ldots,q$, of $A(t)$ the sets

$$\mathfrak{D}_i = \{t \mid \det B_i(t) \neq 0, B_i^{-1}(t)b(t) > 0, d'_{B_i}(t) - c'_{B_i}(t) B_i^{-1}(t) D_{B_i}(t) > 0\} \tag{9}$$

(where $d_{B_i}(t)$ and $D_{B_i}(t)$ correspond to the nonbasic parts of $c(t)$ and $A(t)$, respectively) cover T, i.e.

$$T \subset \bigcup_{i=1}^{q} \mathfrak{D}_i. \tag{10}$$

Then obviously $\gamma(t)$ is continuous on T, since for an arbitrary $t \in T$ we know from

(10) that there is a $\mathfrak{D}_i$ such that $t \in \mathfrak{D}_i$ and $\gamma(t) = c'_{B_i}(t) B_i^{-1}(t) b(t)$ is continuous at t because $\mathfrak{D}_i$ is an open set in $\mathbb{R}^r$.
Now assume that for an arbitrary $t \in \mathfrak{D}_i$ there are vectors $w \in \mathbb{R}^n$, $u \in \mathbb{R}^m$ such that $w \neq 0$, $w \geq 0$, $A(t)w = 0$ and $u \neq 0$, $A'(t)u \leq 0$. Rearranging w into its basic part w_{B_i} and its nonbasic part w_{N_i} we have

$$A(t)w = B_i(t) w_{B_i} + D_{B_i}(t) w_{N_i} = 0$$

and therefore

$$w_{B_i} = -B_i^{-1}(t) D_{B_i}(t) w_{N_i} \geq 0.$$

Obviously $w \neq 0$, $w \geq 0$ implies $w_{N_i} \neq 0$, $w_{N_i} \geq 0$ and therefore

$$\begin{aligned} c'(t)w &= c'_{B_i}(t) w_{B_i} + d'_{B_i}(t) w_{N_i} \\ &= (d'_{B_i}(t) - c'_{B_i}(t) B_i^{-1}(t) D_{B_i}(t)) w_{N_i} > 0 \end{aligned}$$

since $t \in \mathfrak{D}_i$. Furthermore $A'(t)u \leq 0$ is equivalent to

$$\begin{aligned} B'_i(t)u &\leq 0 \\ D'_{B_i}(t)u &\leq 0, \end{aligned}$$

where $u \neq 0$ implies $B'_i(t)u \neq 0$, since det $B_i(t) \neq 0$.
Hence,

$$b'(t)u = b'(t) B_i'^{-1}(t) \cdot B'_i(t)u < 0 \quad \text{for} \quad t \in \mathfrak{D}_i.$$

So we have shown that for (10) the following conditions are necessary, which are a strengthening of (8):

$$\begin{aligned} &w \in \mathbb{R}^n, w \neq 0, w \geq 0, A(t)w = 0 \quad \text{implies} \quad c'(t)w > 0 \quad \text{for all} \quad t \in T \\ &u \in \mathbb{R}^m, u \neq 0, A'(t)u \leq 0 \quad \text{implies} \quad b'(t)u < 0 \quad \text{for all} \quad t \in T. \end{aligned} \tag{11}$$

We cannot expect that (11) implies (10), but it can be proved, that (11) implies the continuity of $\gamma(t)$ on a compact interval T.

Lemma 13. *Let $T \subset \mathbb{R}^r$ be a compact interval. Given* (11), *the set*

$$\mathfrak{B} = \{(x,u) \mid \exists t \in T: A(t)x = b(t), x \geq 0, c'(t)x - b'(t)u \leq 0, A'(t)u \leq c(t)\}$$

is a bounded subset of $\mathbb{R}^n \times \mathbb{R}^m$.

Proof. Assume that $\mathfrak{B}$ is not bounded. Then there is a sequence $\{(x_\nu, u_\nu)\}$ in $\mathfrak{B}$ such that the Euclidean norm $\|(x_\nu, u_\nu)\| \geq \nu$, $\nu = 1, 2, 3, \ldots$. Since T is compact, this sequence may be chosen so that the corresponding $t_\nu \in T$ converge to some $t^* \in T$. From $\|(x_\nu, u_\nu)\| \leq \|x_\nu\| + \|u_\nu\|$ it follows that at least one of the sequences $\{x_\nu\}$ and $\{u_\nu\}$ is unbounded.
If the sequence $\{x_\nu\}$ is unbounded, we may take a subsequence $\{x_{\nu_i}\}$ such that $\|x_{\nu_i}\| \geq i$ for all ν_i, $i = 1, 2, \ldots$, and such that

$$\xi_i = \frac{x_{\nu_i}}{\|x_{\nu_i}\|} \quad \text{converges to } \xi^*.$$

Obviously $\xi^* \geq 0$ and $\|\xi^*\| = 1$.

Since $A(t), b(t), c(t)$ are, due to (5), continuous in t, we have

$$\lim_{i\to\infty} A(t_{\nu_i})\xi_i = A(t^*)\xi^* = \lim_{i\to\infty} \frac{b(t_{\nu_i})}{\|x_{\nu_i}\|} = 0.$$

Then from (11) follows that

$$\lim_{i\to\infty} c'(t_{\nu_i})\xi_i = c'(t^*)\xi^* > 0$$

and therefore

(12) $$c'(t_{\nu_i})x_{\nu_i} = \|x_{\nu_i}\| \cdot c'(t_{\nu_i})\xi_i \to +\infty.$$

If the sequence $\{u_\nu\}$ is unbounded, we choose a subsequence $\{u_{\kappa_j}\}$ such that $\|u_{\kappa_j}\| \geq j$ for all $\kappa_j, j = 1, 2, \ldots$, and at the same time $\eta_j = \frac{u_{\kappa_j}}{\|u_{\kappa_j}\|}$ converges to η^*. Obviously $\|\eta^*\| = 1$ and

$$A'(t^*)\eta^* = \lim_{j\to\infty} A'(t_{\kappa_j})\eta_j \leq \lim_{j\to\infty} \frac{c(t_{\kappa_j})}{\|u_{\kappa_j}\|} = 0$$

which yields, according to (11),

$$\lim_{j\to\infty} b'(t_{\kappa_j})\eta_j = b'(t^*)\eta^* < 0$$

and therefore

(13) $$b'(t_{\kappa_j})u_{\kappa_j} = \|u_{\kappa_j}\| b'(t_{\kappa_j})\eta_j \to -\infty.$$

But (12) as well as (13) is a contradiction to

(14) $$c'(t_\nu)x_\nu - b'(t_\nu)u_\nu \leq 0, \qquad \nu = 1, 2, 3, \ldots,$$

which has to be satisfied for $(x_\nu, u_\nu) \in \mathfrak{B}$. Hence, our assumption, that $\mathfrak{B}$ is not bounded, is contradictory. □

Theorem 14. *If $T \subset \mathbb{R}^r$ is a compact interval and if* (11) *is valid on T, then $\gamma(t)$ is continuous on T.*

Proof. For any $t \in T$ the vectors $x^* \in \mathbb{R}^n$ and $u^* \in \mathbb{R}^m$ are solutions of (7) and its dual program if and only if $x^* \geq 0$ and, for all $u \in \mathbb{R}^m$ and all $x \in \mathbb{R}^n$ such that $x \geq 0$,

(15) $$\begin{aligned} c'(t)x^* + u'(b(t) - A(t)x^*) &\leq c'(t)x^* + u^{*\prime}(b(t) - A(t)x^*) \leq \\ &\leq c'(t)x + u^{*\prime}(b(t) - A(t)x) \end{aligned}$$

(see Th. 0.13 (Kuhn-Tucker)), where $b(t) - A(t)x^* = 0$.
Hence, if $\hat{t} \in T$ and $\bar{t} \in T$ and $\hat{x}, \hat{u}$ and $\bar{x}, \bar{u}$ are primal and dual solutions corresponding to $\hat{t}$ and $\bar{t}$ respectively, we conclude from (15) that

(16) $$\begin{aligned} |(\gamma(\hat{t}) - \gamma(\bar{t}))| = |c'(\hat{t})\hat{x} - c'(\bar{t})\bar{x}| \leq \max\{&|\Delta c'\bar{x} + \hat{u}'(\Delta b - \Delta A\bar{x});\\ &|\Delta c'\hat{x} + \bar{u}'(\Delta b - \Delta A\hat{x})|\} \end{aligned}$$

where

$$\Delta c = c(\hat{t}) - c(\bar{t})$$
$$\Delta b = b(\hat{t}) - b(\bar{t})$$
$$\Delta A = A(\hat{t}) - A(\bar{t}).$$

Since according to the duality theorem (see Th. 0.9)$(\bar{x},\bar{u})$ and $(\hat{x},\hat{u})$ are elements of the bounded set $\mathfrak{B}$ defined in Lemma 13, the continuity of $\gamma(t)$ on t follows from (16). □

This result suggests the application of numerical quadrature, if we are interested in determining $F_{\gamma(t)}(\xi)$ or $E\gamma^{\nu}(t)$ and t is a random vector with range in the compact interval T and with a continuous density function $f_t(\tau)$. However, this leads in general to a tremendeous amount of work. The question, whether numerical quadrature or Monte-Carlo simulation is to be preferred, has not yet been answered.

Chapter III. Two Stage Problems

1. The General Case

In Ch. I we introduced the general two-stage problem of SLP as

$$\text{(1)} \qquad \min_{P_x \in \mathfrak{P}} F(E_{P_x \times P_\omega} L^i(e(\omega,x)); \quad i=1,\ldots,r), \quad \text{where}$$

$$L(e(\omega,x)) = c'(\omega)x + \min\{q'y \mid Wy = b(\omega) - A(\omega)x, \, y \geq 0\}.$$

As stated there, q and W may also be stochastic in the sense that they are random vectors on the probability space $(\Omega, \mathfrak{F}, P_\omega)$ too. First of all we shall check whether we really need mixed strategies or not. In many practical cases, the problem will be stated somewhat different from (1), namely

$$\text{(2)} \qquad \min_{P_x \in \mathfrak{P}} F(E_{P_x \times P_\omega} \tilde{L}^i(e(\omega,x)); i=1,\ldots,r), \quad \text{where}$$

$$\tilde{L}^i(e(\omega,x)) = \begin{cases} L(e(\omega,x)), & \text{if} \quad i=1 \\ |L^i(e(\omega,x))|, & \text{if} \quad i>1. \end{cases}$$

The question, whether we can restrict ourselves to pure strategies without loss, was first answered by Wessels [18] for the case when

$$F(E_{P_x \times P_\omega} \tilde{L}^i(e(\omega, x)); \quad i=1,\ldots,r)$$
$$= \mu E_{P_x \times P_\omega} L(e(\omega, x)) + \lambda \sigma_{P_x \times P_\omega} L(e(\omega,x)),$$

where $\lambda \geq 0$ and σ means the standard deviation

$$\sigma_{P_x \times P_\omega} = \sqrt{E_{P_x \times P_\omega} \{L(e(\omega,x)) - E_{P_x \times P_\omega} L(e(\omega,x))\}^2}.$$

This has been extended by Marti [11] to the case when

$$F(E_{P_x \times P_\omega} \tilde{L}^i(e(\omega,x)); \quad i=1,\ldots,r)$$
$$= \sum_{i=1}^{r} \lambda_i \sqrt[i]{E_{P_x \times P_\omega} \tilde{L}^i(e(\omega,x))}, \qquad \lambda_i \geq 0.$$

The basis for these statements is the following

Theorem 1. *Suppose that*

$$\sqrt[i]{\int_{X \times \Omega} \tilde{L}^i(e(\omega,x)) \mathrm{d}(P_x \times P_\omega)} \quad \textit{exists.}$$

Then

$$\inf_{x \in X} \{\int_\Omega \tilde{L}^i(e(\omega,x)) \mathrm{d}P_\omega\}^{\frac{1}{i}} \leq \{\int_{X \times \Omega} \tilde{L}_i(e(\omega,x)) \mathrm{d}(P_x \times P_\omega)\}^{\frac{1}{i}}$$

Proof. First let $i=1$; then

$$\inf_{x\in X}\int_\Omega L(e(\omega,x))\,\mathrm{d}P_\omega \le \int_\Omega L(e(\omega,x))\,\mathrm{d}P_\omega \quad \forall x\in X \quad \text{and}$$

hence, for any P_x with $P_x(X)=1$,

$$\begin{aligned}\inf_X \int_\Omega L(e(\omega,x))\,\mathrm{d}P_\omega &= \int_X \{\inf_X \int_\Omega L(e(\omega,x))\,\mathrm{d}P_\omega\}\,\mathrm{d}P_x \le \\ &\le \int_X \int_\Omega L(e(\omega,x))\,\mathrm{d}P_\omega \,\mathrm{d}P_x \\ &\int_{X\times\Omega} L(e(\omega,x))\,\mathrm{d}(P_\omega\times P_x),\end{aligned}$$

by Fubini's theorem.

Now let $i>1$ and define

$$G(x)=\left\{\int_\Omega \tilde{L}^i(e(\omega,x))\,\mathrm{d}P_\omega\right\}^{\frac{1}{i}}.$$

Then by Hölder's inequality and Fubini's theorem

$$\begin{aligned}\inf_X \; G(x) &= \int_X \inf_x G(x)\,\mathrm{d}P_x \le \int_X G(x)\,\mathrm{d}P_x \\ &\le \left\{\int_X G^i(x)\,\mathrm{d}P_x\right\}^{\frac{1}{i}} \cdot \left\{\int_X \mathrm{d}P_x\right\}^{\frac{1}{j}}; \quad \frac{1}{i}+\frac{1}{j}=1 \\ &= \left\{\int_X\int_\Omega \tilde{L}^i(e(\omega,x))\,\mathrm{d}P_\omega\,\mathrm{d}P_x\right\}^{\frac{1}{i}} \\ &= \left\{\int_{X\times\Omega} \tilde{L}^i(e(\omega,x))\,\mathrm{d}(P_\omega\times P_x)\right\}^{\frac{1}{i}},\end{aligned}$$

which completes the proof. □

Corollary 2. *If* $F(E_{P_x\times P_\omega}\tilde{L}^i(e(\omega,x));\, i=1,\dots,r)=\sum_{i=1}^r \lambda_i\{E_{P_x\times P_\omega}\tilde{L}^i(e(\omega,x))\}^{\frac{1}{i}}$, *where* $\lambda_i\ge 0$ for $i=1,\dots,r$, *and all integrals involved are supposed to exist, then, for any* P_x *with* $P_x(X)=1$,

$$\inf_{x\in X}\sum_{i=1}^r \lambda_i\{E_{P_\omega}\tilde{L}^i(e(\omega,x))\}^{\frac{1}{i}} \le \sum_{i=1}^r \lambda_i\{E_{P_\omega\times P_x}\tilde{L}^i(e(\omega,x))\}^{\frac{1}{i}}.$$

Proof. Obviously

$$\inf_{x\in X}\sum_{i=1}^r \lambda_i\{E_{P_\omega}\tilde{L}^i(e(\omega,x))\}^{\frac{1}{i}} \le \sum_{i=1}^r \lambda_i \int_X\left\{\int_\Omega \tilde{L}^i(e(\omega,x))\,\mathrm{d}P_\omega\right\}^{\frac{1}{i}}\mathrm{d}P_x.$$

Applying Fubini's theorem and Hölder's inequality in the same way as in the proof of Th. 1 yields the desired result. □

For problem (1) a similar result may easily be established:

Theorem 3. *Provided the integrals involved exist*

$$\inf_X \sum_{i=1}^r \lambda_i\{E_{P_\omega} L^i(e(\omega,x))\} \le \sum_{i=1}^r \lambda_i\{E_{P_\omega\times P_x} L^i(e(\omega,x))\}$$

for any P_x *with* $P_x(X)=1$.

Proof. $\inf_X \sum_{i=1}^{r} \lambda_i \int_{\Omega} L^i(e(\omega,x))\mathrm{d}P_\omega = \int_X \left\{ \inf_X \int_{\Omega} \sum_{i=1}^{r} \lambda_i L^i(e(\omega,x))\mathrm{d}P_\omega \right\} \mathrm{d}P_x$

$$\leq \int_X \int_{\Omega} \sum_{i=1}^{r} \lambda_i L^i(e(\omega,x))\, \mathrm{d}P_\omega \mathrm{d}P_x$$

$$= \sum_{i=1}^{r} \lambda_i \int_{X\times\Omega} L^i(e(\omega,x))\mathrm{d}(P_\omega \times P_x). \; \square$$

The three statements Th. 1, Cor. 2 and Th. 3 cover more than we shall treat in this book. However, the type of function — F in (1) as well as in (2) — with which we may restrict ourselves without loss to pure strategies, does not, in general, seem to be answered.
Furthermore, in the following we shall reduce problem (1), with the help of our previous results, to

(3) $$\min_{x\in X} E_{P_\omega} \{c'(\omega)x + \min_y \{q'(\omega)y \mid W(\omega)y = b(\omega) - A(\omega)x,\, y \geq 0\}\}.$$

First we have to discuss what we call the feasible set of decisions. Here we are especially concerned with the so-called recourse program

(4) $$Q(x,\omega) = \begin{cases} \inf\{q'(\omega)y \mid W(\omega)y = b(\omega) - A(\omega)x,\, y \geq 0\}, & \text{if feasible } y \text{ exist;} \\ +\infty, & \text{if no feasible } y \text{ exist.} \end{cases}$$

Since we are dealing with programming with recourse, it would be meaningless to allow "here and now" decisions x for which in the second stage problem (4) recourse is impossible, i.e. $Q(x,\omega) = +\infty$, with a positive probability. Hence we define

(5) $$K = \{x \mid x \in \mathbb{R}^n; Q(x,\omega) < +\infty \text{ with probability } 1\}.$$

Obviously $x \in K$ does not yet establish the existence of the integral involved in (3); it is only a necessary condition. Furthermore we still require $x \in X$, where $X \subset \mathbb{R}^n$ is some predetermined set, usually given by simultanuous linear constraints. Therefore we shall assume throughout that X is a convex polyhedral set. Concerning K, Walkup and Wets [15] found the following result:

Theorem 4. *K is a closed convex set.*

Proof. First we show the convexity of K.
Let $x_1 \in K$ and $x_2 \in K$.
Define

$$\Omega_1 = \{\omega \mid Q(x_1,\omega) < +\infty\}$$
$$\Omega_2 = \{\omega \mid Q(x_2,\omega) < +\infty\}.$$

Then

$$P_\omega(\Omega_1) = P_\omega(\Omega_2) = 1, \quad \text{which implies}$$
$$\begin{aligned} P_\omega(\Omega_1 \cap \Omega_2) &= P_\omega(\Omega_1 - (\Omega_1 - \Omega_2)) \\ &= P_\omega(\Omega_1) - P_\omega(\Omega_1 - \Omega_2) \\ &\geq P_\omega(\Omega_1) - P_\omega(\Omega - \Omega_2) \\ &= P_\omega(\Omega_1) - [P_\omega(\Omega) - P_\omega(\Omega_2)] = 1 \end{aligned}$$

and hence

$$P_\omega(\Omega_1 \cap \Omega_2) = 1.$$

For any $\omega \in \Omega_1 \cap \Omega_2$ there exist y_1 and y_2 such that

$$W(\omega)y_1 = b(\omega) - A(\omega)x_1, \quad y_1 \geq 0$$
$$W(\omega)y_2 = b(\omega) - A(\omega)x_2, \quad y_2 \geq 0.$$

Therefore, for any $\lambda \in (0,1)$

$$W(\omega)\,[\lambda y_1 + (1-\lambda)y_2] = b(\omega) - A(\omega)\,[\lambda x_1 + (1-\lambda)x_2]$$
$$\text{and} \quad \lambda y_1 + (1-\lambda)y_2 \geq 0.$$

Hence,

$$\Omega_\lambda = \{\omega \mid Q(\lambda x_1 + (1-\lambda)x_2; \omega) < +\infty\} \supset \Omega_1 \cap \Omega_2,$$

which means that

$$P_\omega(\Omega_\lambda) = 1.$$

This establishes the convexity of K.
Now we have to show that K is closed.
Let $\{x_i; i = 1, 2, \ldots\}$ be a sequence in K converging to $\hat{x}$. Then for

$$\Omega_i = \{\omega \mid Q(x_i, \omega) < +\infty\}$$

we have

$$P_\omega(\Omega_i) = 1.$$

Let

$$\Lambda_k = \bigcap_{i=1}^{k} \Omega_i.$$

Obviously we have

$$\Lambda_{k+1} \subset \Lambda_k,$$

and hence

$$\lim_{k\to\infty} \Lambda_k = \bigcap_{k=1}^{\infty} \Lambda_k = \bigcap_{i=1}^{\infty} \Omega_i \quad \text{and}$$

by induction from the first part of the proof

$$P_\omega(\Lambda_k) = 1; \quad k = 1, 2, \ldots.$$

and hence

$$P_\omega\left(\bigcap_{i=1}^{\infty} \Omega_i\right) = P_\omega\left(\lim_{k\to\infty} \Lambda_k\right) = \lim_{k\to\infty} P_\omega(\Lambda_k) = 1.$$

Suppose now $\hat{x} \notin K$.

Then there is an

$$\omega \in \bigcap_{i=1}^{\infty} \Omega_i,$$

such that, for the Euclidean norm,

$$\| W(\omega)y - b(\omega) - A(\omega)\hat{x} \| \geq \varrho > 0 \quad \forall y \geq 0.$$

Then

$$U_{\hat{x}} = \{x \mid \| W(\omega)y - b(\omega) - A(\omega)x \| \geq \frac{\varrho}{2} \; \forall y \geq 0\}$$

is a neighbourhood of $\hat{x}$; since $\{x_i\}$ converges to $\hat{x}$, there must be an $x_k \in U_{\hat{x}}$. For this x_k

$$Q(x_k, \omega) < +\infty, \quad \text{since} \quad \omega \in \bigcap_{i=1}^{\infty} \Omega_i \subset \Omega_k,$$

which yields the existence of a $y_k \geq 0$ such that

$$W(\omega)y_k = b(\omega) - A(\omega)x_k$$

in contradiction to $x_k \in U_{\hat{x}}$.
From this we must conclude that $\hat{x} \in K$, which means that K is closed. □

Now we shall investigate the properties of the objectives involved in (3). First we state

Theorem 5. *$Q(x, \omega)$ is, almost surely on Ω, a convex function on K.*

Proof. Let $x_1 \in K$ and $x_2 \in K$ and $\Omega_i = \{\omega \mid Q(x_i, \omega) < +\infty\}$, $i = 1, 2$. As we know from the proof of Th. 4

$$P_\omega(\Omega_1 \cap \Omega_2) = 1.$$

Hence it suffices to show that

$$Q(\lambda x_1 + (1-\lambda)x_2; \omega) \leq \lambda Q(x_1, \omega) + (1-\lambda) Q(x_2, \omega) \;\; \forall \omega \in \Omega_1 \cap \Omega_2 \;\; \text{and} \;\; \forall \lambda \in (0,1).$$

Here

$$(6) \qquad \begin{aligned} Q(x_i, \omega) = \inf q'(\omega) y & \\ \text{subject to} \quad W(\omega)y = b(\omega) - A(\omega)x_i; & \quad i = 1, 2, \\ y \geq 0 & \end{aligned}$$

and

$$(7) \qquad \begin{aligned} Q(\lambda x_1 + (1-\lambda)x_2; \omega) = \inf q'(\omega) y & \\ \text{subject to} \quad W(\omega)y = b(\omega) - A(\omega)(\lambda x_1 + (1-\lambda)x_2) & \\ y \geq 0. & \end{aligned}$$

Let $y_i, i = 1, 2$, be feasible in (6). Then obviously $\lambda y_1 + (1-\lambda)y_2$ is feasible in (7) for any $\lambda \in (0,1)$.
If $Q(x_i, \omega)$, $i = 1, 2$, are both finite, then we can choose y_i to be the solutions of (6) respectively. Hence,

$$Q(\lambda x_1+(1-\lambda)x_2,\omega)\le q'(\omega)(\lambda y_1+(1-\lambda)y_2)=\lambda Q(x_1,\omega)+(1-\lambda)Q(x_2,\omega).$$

If $Q(x_1,\omega)=-\infty$, we know that there exists a $y^*\ge 0$ such that

$$W(\omega)y^*=0 \quad \text{and} \quad q'(\omega)y^*<0$$

which implies $Q(\lambda x_1+(1-\lambda)x_2;\omega)=-\infty$, since

$$\lambda y_1+(1-\lambda)y_2+\mu y^* \quad \text{is feasible in (7) for any} \quad \mu>0,$$

and $q'(\omega)\{\lambda y_1+(1-\lambda)y_2+\mu y^*\}$ tends to $-\infty$ as $\mu\to+\infty$. Therefore, we may in this case also state

$$Q(\lambda x_1+(1-\lambda)x_2;\omega)\le\lambda Q(x_1,\omega)+(1-\lambda)Q(x_2,\omega). \ \square$$

After this theorem we may discuss

$$Q(x)=\int_\Omega Q(x,\omega)\mathrm{d}P_\omega.$$

We may rewrite this integral as

$$(8) \qquad Q(x)=\int_\Omega Q^+(x,\omega)\mathrm{d}P_\omega-\int_\Omega Q^-(x,\omega)\mathrm{d}P_\omega=Q^+(x)-Q^-(x)$$

where

$$Q^+(x,\omega)=\begin{cases}Q(x,\omega) & \text{whenever} \quad Q(x,\omega)>0\\ 0 & \text{otherwise}\end{cases}$$

and

$$Q^-(x,\omega)=\begin{cases}-Q(x,\omega) & \text{whenever} \quad Q(x,\omega)<0\\ 0 & \text{otherwise.}\end{cases}$$

In spite of the fact, that the restriction $x\in K$ implies

$$P_\omega(\{\omega\,|\,Q^+(x,\omega)=+\infty\})=0,$$

it is still possible that $Q^+(x)=+\infty$. This means that either $Q(x)$ would be undefined, if $Q^-(x)=+\infty$, or $Q(x)=+\infty$. Since both these situations are meaningless in practice, because one does not want decisions with either an undefined outcome or infinitely high costs, it seems natural to restrict x to

$$\tilde{K}=\{x\,|\,Q^+(x)<+\infty\}.$$

Obviously $\tilde{K}\subset K$. Since $Q^-(x)=+\infty$ is not meaningful in the practical interpretation of the recourse program, which was introduced to compensate for some inconvenient situation caused by the original decision on x and the realization of ω usually implying additional (positive) costs, we need not consider it in the following

Theorem 6. *$Q(x)$ is a convex function on $\tilde{K}$, and hence $\tilde{K}$ is convex.*

Proof. Since $x \in \tilde{K}$, $Q(x) < +\infty$. Let $x_1 \in \tilde{K}$ and $x_2 \in \tilde{K}$, implying $x_i \in K$, $i=1,2$. Due to Th. 5 for any $\lambda \in (0,1)$ $Q(\lambda x_1 + (1-\lambda)x_2; \omega) \leq \lambda Q(x_1, \omega) + (1-\lambda)Q(x_2, \omega)$ almost surely.
Hence

$$\begin{aligned} Q(\lambda x_1 + (1-\lambda)x_2) &= \int_\Omega Q(\lambda x_1 + (1-\lambda)x_2, \omega)\, dP_\omega \leq \\ &\leq \int_\Omega \{\lambda Q(x_1 \omega) + (1-\lambda)Q(x_2, \omega)\}\, dP_\omega \\ &= \lambda Q(x_1) + (1-\lambda)Q(x_2). \end{aligned}$$

This inequality also establishes the convexity of $\tilde{K}$. □

Corollary 7. *Either $Q(x) > -\infty$ on $\tilde{K}$ or $Q(x) = -\infty$ on the relative interior of $\tilde{K}$.*

Proof. Follows immediately from the convexity of $Q(x)$ on $\tilde{K}$. □

However, as was pointed out by Walkup and Wets [15], $\tilde{K}$ need not be closed in general, and $Q(x)$ may be discontinuous on $\tilde{K}$, i.e. at the relative boundary, even if $Q(x)$ is finite on $\tilde{K}$.

2. The Fixed Recourse Case

We are now considering two-stage problems where

$$W(\omega) \equiv W$$

i.e. W is a fixed nonstochastic matrix. From the previous section we know that in general $\tilde{K} \subset K$. Under suitable integrability assumptions on the original random variables we may strengthen this statement in the fixed recourse case. The following four statements are due to Walkup and Wets [15].

Theorem 8. *Let the random variables in $q(\omega)$, $b(\omega)$, $A(\omega)$ be square integrable. Then $\tilde{K} = K$.*

Proof: We have to show that $K \subset \tilde{K}$. Let $\hat{x} \in K$ be arbitrarily chosen. For $\Omega(\hat{x}) = \{\omega \mid Q(\hat{x}, \omega) < +\infty\}$ we know by the definition of K, that $P_\omega(\Omega(\hat{x})) = 1$. If we define $\Omega_1 = \{\omega \mid \omega \in \Omega(\hat{x}), |Q(\hat{x}, \omega)| < \infty\}$, then obviously $Q^+(\hat{x}) \leq \int_{\Omega_1} |Q(\hat{x}, \omega)|\, dP_\omega$.

Since $Q(\hat{x}, \omega)$ is finite on Ω_1, we may represent $Q(\hat{x}, \omega)$ in terms of basic solutions for any $\omega \in \Omega_1$, i.e. $Q(\hat{x}, \omega) = \tilde{q}'(\omega) B^{-1}[b(\omega) - A(\omega)\hat{x}]$, where B is an optimal feasible basis out of W with respect to ω and $\tilde{q}(\omega)$ is the vector of the components of $q(\omega)$ corresponding to B. By the assumed square integrability of the elements of $q(\omega)$, $b(\omega)$ and $A(\omega)$ and Schwarz's inequality it is now obvious that $|Q(\hat{x}, \omega)|$ is integrable on Ω_1.
Hence

$$Q^+(\hat{x}) < \infty \quad \text{implying} \quad \hat{x} \in \tilde{K}. \ \square$$

Also Cor. 7 may be strengthened to

Corollary 9. *Under the assumption of Th. 8 either*

$$Q(x) > -\infty \quad on \quad K \quad or$$
$$Q(x) = -\infty \quad on \quad K.$$

Proof. We have to show, that $Q(x)=-\infty$ for some $x \in K$ implies $Q(\bar{x})=-\infty$ for every other point $\bar{x} \in K$.
Let x be a point of K so that $Q(x)=-\infty$.
From our integrability assumption it follows that $P_\omega(\{\omega \mid Q(x,\omega)=-\infty\})=\alpha>0$.
Let $\Omega_\infty=\{\omega \mid \exists y \geq 0: Wy=0, q'(\omega)y<0\}$ (see Th. 0.4).
Then

$$P_\omega(\Omega_\infty)=\alpha.$$

If $\bar{x}$ is any other point of K, then for

$$\Omega(\bar{x})=\{\omega \mid Q(\bar{x},\omega)<+\infty\}$$

we know that

$$P_\omega(\Omega(\bar{x}))=1.$$

Hence,

$$P_\omega(\Omega_\infty \cap \Omega(\bar{x}))=P_\omega(\Omega_\infty)-P_\omega(\Omega_\infty \cap (\Omega-\Omega(\bar{x})))$$
$$=\alpha$$

since $P_\omega(\Omega-\Omega(\bar{x}))=0$.
For $\omega \in \Omega_\infty \cap \Omega(\bar{x})$ there exists a feasible solution to the recourse program — since $\omega \in \Omega(\bar{x})$ — and a direction $y \geq 0$ such that $Wy=0$ and $q'(\omega)y<0$ — since $\omega \in \Omega_\infty$ — which implies

$$Q(\bar{x},\omega)=-\infty.$$

Hence $P_\omega(\{\omega \mid Q(\bar{x},\omega)=-\infty\}) \geq \alpha > 0$ and therefore

$$Q(\bar{x})=-\infty. \; \square$$

Theorem 10. *Let the random variables in $q(\omega)$, $b(\omega)$, $A(\omega)$ be square integrable and $Q(x)>-\infty$ on K. Then $Q(x)$ is Lipschitz continuous.*

Proof. Let $x_i \in K$, $i=1,2$ and $\Omega_i=\{\omega \mid Q(x_i,\omega)<+\infty\}$, hence $P_\omega(\Omega_i)=1$, $i=1,2$.
From $Q(x)>-\infty$ on K it follows, for

$$\Omega_\infty=\{\omega \mid \exists y \geq 0: Wy=0, q'(\omega)y<0\},$$

that $P_\omega(\Omega_\infty)=0$.
Therefore $P_\omega((\Omega_1 \cap \Omega_2)-\Omega_\infty)=1$.
For any $\omega \in (\Omega_1 \cap \Omega_2)-\Omega_\infty$ we have $-\infty<Q(x_i,\omega)<+\infty$
and, obviously, $-\infty<Q(\lambda x_1+(1-\lambda)x_2,\omega)<+\infty$.
Representing the optimal value via basic solutions shows that $Q(x,\omega)$ is piecewise linear on

$\{x \mid x=\lambda x_1+(1-\lambda)x_2, \lambda \in [0,1]\}$ and, due to Th. 5, convex.

Hence

$$Q(x_i, \omega) = \alpha_{v_i}(\omega) + d'_{v_i}(\omega) x_i,$$

(see Th. 0.6), where $\alpha_{v_i}(\omega)$ is a weighted sum of products of elements of $q(\omega)$ and $b(\omega)$ and $d_{v_i}(\omega)$ consists of components, which are weighted sums of elements of $q(\omega)$ and $A(\omega)$. Hence, according to our square integrability assumption and Schwarz's inequality, $\alpha_{v_i}(\omega)$ and $d_{v_i}(\omega)$ are integrable. If $0 \leq Q(x_2, \omega) - Q(x_1, \omega)$, then from the convexity of $Q(x, \omega)$ in x we have (see Th. 0.11)

$$Q(x_2, \omega) + (x_1 - x_2)' d_{v_2}(\omega) \leq Q(x_1, \omega),$$

implying

$$\alpha_{v_2}(\omega) - \alpha_{v_1}(\omega) \leq \left(d_{v_1}(\omega) - d_{v_2}(\omega)\right)' x_1.$$

Hence,

$$\begin{aligned} |Q(x_2, \omega) - Q(x_1, \omega)| &= \alpha_{v_2}(\omega) - \alpha_{v_1}(\omega) + d'_{v_2}(\omega) x_2 - d'_{v_1}(\omega) x_1 \\ &\leq d'_{v_2}(\omega)(x_2 - x_1). \end{aligned}$$

If

$$Q(x_2, \omega) - Q(x_1, \omega) < 0,$$

we get in an analogous way

$$|Q(x_2, \omega) - Q(x_1, \omega)| \leq d'_{v_1}(\omega)(x_1 - x_2).$$

Hence we have in general

$$|Q(x_2, \omega) - Q(x_1, \omega)| \leq \operatorname*{Max}_i |d'_{v_i}(\omega)(x_2 - x_1)|.$$

Since for any two vectors $f, g \in \mathbb{R}^n$

$$|f'g| \leq \sqrt{n} |f|_\infty \cdot \|g\|,$$

where $|\cdot|_\infty$ is the maximum norm and $\|\cdot\|$ indicates the Euclidean norm, we have

$$|Q(x_2, \omega) - Q(x_1, \omega)| < \sqrt{n} \operatorname*{Max}_i |d_{v_i}(\omega)|_\infty \cdot \|x_2 - x_1\|.$$

Since obviously $\operatorname*{Max}_i |d_{v_i}(\omega)|_\infty$ is integrable on $(\Omega_1 \cap \Omega_2) - \Omega_\infty$, and the same is true for
$\operatorname*{Max}_v |d_v(\omega)|_\infty$, where v varies over the number of all possible bases, we get

$$\begin{aligned} |Q(x_2) - Q(x_1)| &\leq \int_{(\Omega_1 \cap \Omega_2) - \Omega_\infty} |Q(x_2, \omega) - Q(x_1, \omega)| \, dP_\omega \\ &\leq \sqrt{n} \left\{ \int_\Omega \operatorname*{Max}_v |d_v(\omega)| \, dP_\omega \right\} \cdot \|x_2 - x_1\|. \end{aligned}$$

Since $\int_\Omega \operatorname*{Max}_v |d_v(\omega)| \, dP_\omega$ is independent of x_1, x_2, this is the desired Lipschitz condition. □

Corollary 11. *If either*
a) $q(\omega) \equiv q$ and the random elements of $b(\omega)$ and $A(\omega)$ are integrable or

b) $b(\omega)\equiv b$, $A(\omega)\equiv A$ and the components of $q(\omega)$ are integrable or
c) the ranges of $A(\omega)$, $b(\omega)$, $q(\omega)$ are bounded,
then $\tilde{K}=K$ and either $Q(x)\equiv -\infty$ on K or $Q(x)>-\infty$ on K, and in the latter case, $Q(x)$ is Lipschitz continuous on K.

Proof. Under assumption a) or b), the proofs of Th. 8, Cor. 9 and Th. 10 may be reproduced, since integrability of $Q(x,\omega)$ on K, as required there, follows immediately. Under assumption c), the foregoing propositions may also be proved in the same way, observing that for any $x\in K$ $Q^+(x,\omega)$ is either equal to zero or of the type $\tilde{q}'(\omega)B^{-1}[b(\omega)-A(\omega)x]$, and hence bounded above on Ω, which implies $Q^+(x)<\infty$ on K. Since, in Th. 10, any $d_\nu(\omega)$ is of the form $\tilde{q}'(\omega)B^{-1}A(\omega)$, $\operatorname{Max}_\nu|d_\nu(\omega)|_\infty$ is bounded on Ω; hence we may use in this case

$$\varrho=\sup_\Omega \operatorname{Max}_\nu |d_\nu(\omega)|_\infty\cdot\sqrt{n}$$

as Lipschitz constant. □

Theorem 12. *Suppose one of the integrability assumptions of Th. 10 or Corr. 11 and $Q(x)>-\infty$ on K. Suppose further, that the probability measure P on the Euclidean space $\mathbb{R}^l$ spanned by the elements of A,b,q is absolutely continuous with respect to the Lebesgue measure μ_l on $\mathbb{R}^l$ — i.e. P has a density function —, then $Q(x)$ has a continuous gradient on K.*

Remark. We still assume $(A,b,q)=(A(\omega),b(\omega),q(\omega))$ to be random in spite of the fact that we have omitted the ω for simplicity. The following proof consists of two parts. First we observe that $Q(x,\omega)$, $x\in K$, has a gradient a.e. and that its partial difference quotients are bounded by an integrable function. From Lebesgue's theorem then follows the differentiability of $Q(x)=\int Q(x,\omega)\,dP_\omega$. In the second part we demonstrate continuity by using an explicit presentation for the gradient $\nabla Q(x)$.

Proof. According to our assumptions $Q(x,\omega)$ is finite with probability 1 for any $x\in K$, and may, therefore, be represented via basic solutions as

$$Q(x,\omega)=\tilde{q}_i'B_i^{-1}[b-Ax],$$

where B_i is an optimal feasible basis of W and $\tilde{q}_i$ is the vector of those components of q belonging to B_i.
Hence $Q(x,\omega)$ has a gradient — with respect to x — of the form

$$\nabla_x Q(x,\omega)=-(\tilde{q}_i'B_i^{-1}A)'$$

for all $\omega\in\Omega$ such that A,b,q do not belong to one of a finite number of sets of the type

$$S_{ij}=\{A,b,q\,|\,\tilde{q}_i'B_i^{-1}A\neq\tilde{q}_j'B_j^{-1}A;\quad \tilde{q}_i'B_i^{-1}[b-Ax]=\tilde{q}_j'B_j^{-1}[b-Ax]\};\quad i\neq j.$$

For every A,q define

$$E_{A,q}^{i,j}=\{b\,|\,(A,b,q)\in S_{ij}\}.$$

Now either

$$\tilde{q}_i' B_i^{-1} A = \tilde{q}_j B_j^{-1} A, \text{ implying } \quad E_{A,q}^{i,j} = \emptyset \quad \text{and}$$

hence the Lebesgue measure $\mu(E_{A,q}^{i,j}) = 0$
or

$$\tilde{q}_i B_i^{-1} A \neq \tilde{q}_j' B_j^{-1} A \quad \text{implying} \quad \tilde{q}_i' B_i^{-1} - \tilde{q}_j' B_j^{-1} \neq 0'.$$

Then

$$b \in E_{A,q}^{i,j} \quad \text{if}$$

$$(\tilde{q}_i' B_i^{-1} - \tilde{q}_j' B_j^{-1}) b = (\tilde{q}_i' B_i^{-1} - \tilde{q}_j' B_j^{-1}) A x;$$

this means that $E_{A,q}^{i,j}$ is a hyperplane in the space spanned by the elements of b and hence the Lebesgue measure $\mu(E_{A,q}^{i,j}) = 0$.
From $\mu(E_{A,q}^{i,j}) = 0$ for every A, q it follows, by Th. 0.25, that $\mu_l(S_{ij}) = 0$ and, since P is absolutely continuous with respect to μ_l,

$$P(S_{ij}) = 0.$$

Therefore $Q(x, \omega)$ has for every $x \in K$ a gradient of the form

$$\nabla_x Q(x, \omega) = -(\tilde{q}_i' B_i^{-1} A)'$$

for all ω except a set of P_ω-measure zero.
From the proofs of Th. 10 and Cor. 11 we know that, for

$$x_1 \neq x_2, \qquad x_i \in K,$$

$$\frac{|Q(x_1, \omega) - Q(x_2, \omega)|}{\|x_1 - x_2\|} \leq h(\omega)$$

almost surely, where $h(\omega)$ is integrable. In particular, this inequality is valid for all partial difference quotients. Hence (as a consequence of Th. 0.22) $Q(x)$ has a gradient $\nabla Q(x)$, and

$$\nabla Q(x) = \int \nabla_x Q(x, \omega) \, dP_\omega.$$

The continuity of $\nabla Q(x)$ follows from the following observation:

$$\nabla Q(x) = \sum_i \int_{\mathfrak{B}_i(x)} -(\tilde{q}_i' B_i^{-1} A) \, dP,$$

where

$$\mathfrak{B}_i(x) = \mathfrak{A}_i(x) - \bigcup_{k=1}^{i-1} \mathfrak{A}_k(x)$$

and $\mathfrak{A}_i(x)$ is the "optimality set" of the basis B_i, i.e.

$$\mathfrak{A}_i(x) = \{(A, b, q) \mid B_i^{-1} [b - Ax] \geq 0; \quad q' - \tilde{q}_i' B_i^{-1} W \geq 0\}.$$

Obviously we have to show that the symmetric difference of $\mathfrak{B}_i(x + \Delta x)$ and $\mathfrak{B}_i(x)$ tends to a set of P-measure zero as $\Delta x \to 0$. Looking at this symmetric difference

$$\begin{aligned}
\mathfrak{B}_i(x+\Delta x)\Delta\mathfrak{B}_i(x) &= (\mathfrak{B}_i(x+\Delta x)-\mathfrak{B}_i(x))\cup(\mathfrak{B}_i(x)-\mathfrak{B}_i(x+\Delta x))\\
&= \Big[\big(\mathfrak{A}_i(x+\Delta x)-\bigcup_{k=1}^{i-1}\mathfrak{A}_k(x+\Delta x)\big)-\big(\mathfrak{A}_i(x)-\bigcup_{k=1}^{i-1}\mathfrak{A}_k(x)\big)\Big]\\
&\quad\cup\Big[\big(\mathfrak{A}_i(x)-\bigcup_{k=1}^{i-1}\mathfrak{A}_k(x)\big)-\big(\mathfrak{A}_i(x+\Delta x)-\bigcup_{k=1}^{i-1}\mathfrak{A}_k(x+\Delta x)\big)\Big]\\
&= \Big[\mathfrak{A}_i(x+\Delta x)-\bigcup_{k=1}^{i-1}\mathfrak{A}_k(x+\Delta x)-\mathfrak{A}_i(x)\Big]\\
&\quad\cup\Big[\big(\mathfrak{A}_i(x+\Delta x)-\bigcup_{k=1}^{i-1}\mathfrak{A}_k(x+\Delta x)\big)\cap\bigcup_{k=1}^{i-1}\mathfrak{A}_k(x)\Big]\\
&\quad\cup\Big[\mathfrak{A}_i(x)-\bigcup_{k=1}^{i-1}\mathfrak{A}_k(x)-\mathfrak{A}_i(x+\Delta x)\Big]\\
&\quad\cup\Big[\big(\mathfrak{A}_i(x)-\bigcup_{k=1}^{i-1}\mathfrak{A}_k(x)\big)\cap\bigcup_{k=1}^{i-1}\mathfrak{A}_k(x+\Delta x)\Big]\\
&\subset[\mathfrak{A}_i(x+\Delta x)-\mathfrak{A}_i(x)]\\
&\quad\cup\Big[\bigcup_{k=1}^{i-1}(\mathfrak{A}_k(x)-\mathfrak{A}_k(x+\Delta x))\Big]\\
&\quad\cup[\mathfrak{A}_i(x)-\mathfrak{A}_i(x+\Delta x)]\\
&\quad\cup\Big[\bigcup_{k=1}^{i-1}(\mathfrak{A}_k(x+\Delta x)-\mathfrak{A}_k(x))\Big]
\end{aligned}$$

shows that it suffices to prove for every index i, that $\mathfrak{A}_i(x+\Delta x)$ tends to a set differing from $\mathfrak{A}_i(x)$ by a set of measure zero, as Δx tends to 0, i.e.

$$P\big[\lim_{\Delta x\to 0}\{(\mathfrak{A}_i(x+\Delta x)-\mathfrak{A}_i(x))\cup(\mathfrak{A}_i(x)-\mathfrak{A}_i(x+\Delta x))\}\big]=0.$$

From

$(A,b,q)\in\mathfrak{A}_i(x+\Delta x)-\mathfrak{A}_i(x)$ it follows that

$$B_i^{-1}(b-Ax)\geq B_i^{-1}A\Delta x$$

and that

$$B_i^{-1}(b-Ax)\ngeq 0.$$

Hence

$$[B_i^{-1}(b-Ax)]_\nu<0$$

for at least one component and, at the same time,

$$[B_i^{-1}(b-Ax)]_\nu\geq[B_i^{-1}A\Delta x]_\nu.$$

It is now obvious, that this element (A,b,q) is not in

$$\lim_{\Delta x\to 0}(\mathfrak{A}_i(x+\Delta x)-\mathfrak{A}_i(x)),$$

which shows that

$$\lim_{\Delta x \to 0} (\mathfrak{A}_i(x+\Delta x) - \mathfrak{A}_i(x)) = \emptyset.$$

If $A, b, q \in \mathfrak{A}_i(x) - \mathfrak{A}_i(x+\Delta x)$, then

$$B_i^{-1}(b - Ax) \geq 0$$

and

$$B_i^{-1}(b - Ax) \ngeq B_i^{-1} A \Delta x$$

implying for at least one component

$$[B_i^{-1} A \Delta x]_\nu > [B_i^{-1}(b - Ax)]_\nu \geq 0; \quad \text{hence} \quad (A, b, q) \in \lim_{\Delta x \to 0} (\mathfrak{A}_i(x) - \mathfrak{A}_i(x+\Delta x))$$

only if

$$[B_i^{-1}(b - Ax)]_\nu = 0, \quad \text{i.e. only if}$$

(A, b) is an element of one of finitely many hyperplanes in the (A, b)-space. Now it is obvious, that

$$\mu_l\{\lim_{\Delta x \to 0} (\mathfrak{A}_i(x) - \mathfrak{A}_i(x+\Delta x))\} = 0.$$

Since P is absolutely continuous with respect to μ_l, we have

$$P[\lim_{\Delta x \to 0} \{(\mathfrak{A}_i(x+\Delta x) - \mathfrak{A}_i(x)) \cup (\mathfrak{A}_i(x) - \mathfrak{A}_i(x+\Delta x))\}] = 0$$

which completes the proof. □

3. Complete Fixed Recourse

Again throughout this section we suppose $W(\omega) \equiv W$. In section III.1 we were concerned with the feasibility sets K and $\tilde{K}$. This means that we have also taken into consideration such cases in which a first decision $x \in X$ may not yield a feasible solution of the recourse problem with some positive probability.
As we have seen in Ch. I, the practical meaning of the recourse problem was the possibility to compensate for deviations from the original constraints, caused by the a priori decision on $x \in X$ and the a posteriori realization of the random elements in $A(\omega)$, $b(\omega)$. Hence, from a practical point of view it often seems meaningful to require that the recourse program has feasible solutions for every $x \in X$ and almost surely with respect to P_ω. From this requirement arises the following

Definition. W is a complete recourse matrix, if

$$\{y \mid Wy = z, y \geq 0\} \neq \emptyset \quad \text{for every} \quad z \in \mathbb{R}^m.$$

Following immediately from this definition every complete recourse matrix W has rank $r(W) = m$. Furthermore it is obvious, that W must have more than m columns W_i, since in the contrary case for $\hat{z} = -\sum_{i=1}^{m} W_i$

$$\{y \mid Wy = \hat{z}, y \geq 0\} \quad \text{would be empty}$$

caused by the fact that from $r(W)=m$ the uniqueness of the solution of $W\hat{y}=\hat{z}, \hat{y}=(-1,\dots,-1)'$ follows.
In that which follows we try to characterize complete recourse matrices.

Lemma 13. *If W is a complete recourse matrix with $m+1$ columns, then every m columns of W are linearly independent.*

Proof. Since $r(W)=m$, let $W_1,\dots,W_m$ be linearly independent.
Suppose now that $W_1,\dots W_{m-1}, W_{m+1}$ are linearly dependent; then there exist α_i such that

$$W_{m+1}=\sum_{i=1}^{m-1}\alpha_i W_i.$$

Since $-W_m\in\mathbb{R}^m$ and W is a complete recourse matrix, there exist $\beta_i\geq 0$ such that

$$\begin{aligned}-W_m&=\sum_{i=1}^{m+1}\beta_i W_i\\&=\sum_{i=1}^{m}\beta_i W_i+\beta_{m+1}\sum_{i=1}^{m-1}\alpha_i W_i\\&=\sum_{i=1}^{m-1}(\beta_i+\beta_{m+1}\alpha_i)\;W_i+\beta_m W_m\end{aligned}$$

and hence

$$\sum_{i=1}^{m-1}(\beta_i+\beta_{m+1}\alpha_i)\,W_i+(1+\beta_m)\,W_m=0,$$

contradicting the linear independence of $W_1,\dots,W_m$, since $\beta_m\geq 0$ and therefore at least $1+\beta_m>0$. □

If we assume the linear independence of $W_1,\dots,W_m$, which is justified by $r(W)=m$, we may state

Theorem 14. *Let W have $m+\bar{n}$ columns ($\bar{n}\geq 1$). W is a complete recourse matrix if and only if*

$$\mathfrak{D}=\{y\mid Wy=0, y\geq 0; y_i>0, i=1,\dots,m\}\neq\emptyset.$$

Proof. The necessity of the condition may be shown as follows:
Let

$$z=\sum_{i=1}^{m}\beta_i W_i, \quad\text{where}\quad \beta_i<0,\quad i=1,\dots,m.$$

Since W is a complete recourse matrix, there exist δ_i such that

$$z=\sum_{i=1}^{m+\bar{n}}\delta_i W_i, \quad\text{where}\quad \delta_i\geq 0,\quad i=1,\dots,m+\bar{n}.$$

Therefore

$$\sum_{i=1}^{m}\beta_i W_i=\sum_{i=1}^{m+n}\delta_i W_i,$$

implying
$$\sum_{i=1}^{m}(\delta_i-\beta_i)W_i+\sum_{i=m+1}^{m+\bar{n}}\delta_i W_i=0,$$

where
$$\delta_i-\beta_i>0,\quad i=1,\ldots,m;\quad \text{but}\quad \delta_i\geq 0,\quad i=m+1,\ldots,m+\bar{n};$$

consequently $\mathfrak{D}\neq\emptyset$.
Suppose now that $\mathfrak{D}\neq\emptyset$. Then there exist numbers

$$\delta_i\geq 0,\quad i=m+1,\ldots,m+\bar{n}$$
$$\alpha_i<0,\quad i=1,\ldots,m$$

such that for $W_{m+\bar{n}+1}=\sum_{i=m+1}^{m+\bar{n}}\delta_i W_i$

$$W_{m+\bar{n}+1}=\sum_{i=m+1}^{m+\bar{n}}\delta_i W_i=\sum_{i=1}^{m}\alpha_i W_i.$$

For any $z\in\mathbb{R}^m$ there is a unique solution of

$$z=\sum_{i=1}^{m}\beta_i W_i,$$

since $W_1,\ldots,W_m$ are linear independent. If $\beta_i\geq 0, i=1,\ldots,m$, we have no further problem. Suppose, therefore, that, for at least one index i, $\beta_i<0$. Without loss of generality we may assume that

$$\frac{\beta_m}{\alpha_m}=\max_{1\leq i\leq m}\frac{\beta_i}{\alpha_i}$$

which has to be strictly positive, since $\alpha_i<0$, $i=1,\ldots,m$, and $\beta_i<0$ for at least one index i.
From the linear independence of $W_1,\ldots,W_m$ and $\alpha_i<0$, $i=1,\ldots,m$, follows the linear independence of $W_1,\ldots,W_{m-1}$, $W_{m+\bar{n}+1}$.
Hence, there is also a unique solution of

$$z=\sum_{i=1}^{m-1}\gamma_i W_i+\gamma_{m+\bar{n}+1}W_{m+\bar{n}+1}.$$

Using

$$W_{m+\bar{n}+1}=\sum_{i=1}^{m}\alpha_i W_i,$$

this implies

$$z=\sum_{i=1}^{m-1}(\gamma_i+\gamma_{m+\bar{n}+1}\cdot\alpha_i)W_i+\gamma_{m+\bar{n}+1}\alpha_m W_m$$

$$=\sum_{i=1}^{m}\beta_i W_i$$

and hence, because of the linear independence of $W_1, \ldots, W_m$,

$$\begin{aligned}\gamma_i + \gamma_{m+\bar{n}+1}\alpha_i &= \beta_i \qquad i=1,\ldots,m-1\\ \gamma_{m+\bar{n}+1}\alpha_m &= \beta_m.\end{aligned}$$

From this system of equations we obtain

$$\gamma_{m+\bar{n}+1} = \frac{\beta_m}{\alpha_m} > 0$$

and

$$\gamma_i = \beta_i - \frac{\beta_m}{\alpha_m}\cdot\alpha_i \geq 0, \qquad i=1,\ldots,m-1$$

since

$$0 < \frac{\beta_m}{\alpha_m} = \max_{1\leq i\leq m} \frac{\beta_i}{\alpha_i} \geq \frac{\beta_j}{\alpha_j} \quad \text{for every} \quad j=1,\ldots,m, \quad \text{and}$$
$$\alpha_j < 0, \quad j=1,\ldots,m.$$

Hence

$$z = \sum_{i=1}^{m-1} \gamma_i W_i + \gamma_{m+\bar{n}+1} \sum_{i=m+1}^{m+\bar{n}} \delta_i W_i$$

where $\gamma_i \geq 0$; $i=1,\ldots,m-1$; $\gamma_{m+\bar{n}+1} > 0$ and $\delta_i \geq 0$, $i=m+1,\ldots,m+\bar{n}$. Since $z \in \mathbb{R}^m$ was arbitrary, this yields the completeness of W. □

From Cor. 9 we know that the expected value $Q(x)$ of the second stage program's optimal value is either $-\infty$ or finite for every $x \in K$, which equals $\mathbb{R}^n$ in the complete recourse case. In practical applications it seems to be meaningful to assume that $Q(x)$ is finite on $\mathbb{R}^n$. A simple condition for this property yields

Theorem 15. *Given complete recourse and one of the integrability conditions of Th.* 10 *or Cor.* 11 *(for example square integrability of the elements of $q(\omega)$, $b(\omega)$, $A(\omega)$), then $Q(x)$ is finite if and only if*

$$\{z \mid W'z \leq q(\omega)\} \neq \emptyset \textit{ with probability } 1.$$

Proof. For an arbitrarily chosen $x \in \mathbb{R}^n$ the second stage program

$$\begin{aligned}Q(x,\omega) = \inf q'(\omega)y&\\ Wy = b(\omega) - A(\omega)x&\\ y \geq 0&\end{aligned}$$

has feasible solutions for every $\omega \in \Omega$ by the completeness of W.
Following the lines of the proofs of Th. 8 and Cor. 9, $Q(x)$ is finite if and only if $Q(x,\omega)$ is finite with probability 1, hence, by the duality theorem, if and only if

$$\{z \mid W'z \leq q(\omega)\} \neq \emptyset \quad \text{with probability } 1 \ \Box$$

Corollary 16. *Given complete recourse, $q(\omega) \equiv q$ (constant) and $A(\omega), b(\omega)$ integrable, then $Q(x)$ is finite if and only if*

$$\{z \mid W'z \leq q\} \neq \emptyset.$$

Proof. Follows immediately from Th. 15.

As we know from Th. 14 for a complete recourse matrix W there exist constants $\alpha_i<0$, $i=1,\dots,m$, and $\delta_i\geq 0$, $i=m+1,\dots,m+\bar{n}$ such that

$$\sum_{i=1}^{m}\alpha_i W_i=\sum_{i=m+1}^{m+\bar{n}}\delta_i W_i, \quad \text{where} \quad W_1,\dots,W_m$$

are supposed to be linearly independent since $r(W)=m$.
With these constants α_i, δ_i we may state

Corollary 17. *Given complete recourse and one of the integrability conditions assumed in Th. 15, for $Q(x)$ to be finite it is necessary that*

$$\sum_{i=1}^{m}\alpha_i q_i(\omega)\leq \sum_{i=m+1}^{m+\bar{n}}\delta_i q_i(\omega) \quad \textit{with probability } 1.$$

If $\bar{n}=1$ this condition is also sufficient.

Proof. From Th. 15 we know that $Q(x)$ is finite only if $\{z|\,W'z\leq q(\omega)\}\neq\emptyset$ with probability 1, and hence, by Farkas' lemma, only if $\forall u\geq 0$, $Wu=0$ implies $q'(\omega)u\geq 0$ with probability 1. In particular, for $u^*=(-\alpha_1,\dots,-\alpha_m,\delta_{m+1},\dots,\delta_{m+\bar{n}})'\geq 0$, $Wu^*=0$.
Hence $Q(x)$ is finite only if

$$\sum_{i=1}^{m}\alpha_i q_i(\omega)\leq \sum_{i=m+1}^{m+\bar{n}}\delta_i q_i(\omega) \quad \text{with probability 1.}$$

Suppose now that $\bar{n}=1$ and hence

$$\delta_{m+1}W_{m+1}=\sum_{i=1}^{m}\alpha_i W_i$$

where $\alpha_i<0$, $i=1,\dots,m$ and $\delta_{m+1}\geq 0$, implying $\delta_{m+1}>0$ by the linear independence of $W_1,\dots,W_m$, and suppose further that

$$P_\omega\left[\{\omega|\,\sum_{i=1}^{m}\alpha_i q_i(\omega)\leq\delta_{m+1}q_{m+1}(\omega)\}\right]=1.$$

For almost every $\omega\in\Omega$ there exists a unique $z(\omega)$ such that

$$W_i'z(\omega)=q_i(\omega), \quad i=1,\dots,m, \quad \text{which implies}$$

$$\delta_{m+1}W_{m+1}'z(\omega)=\sum_{i=1}^{m}\alpha_i W_i'z(\omega)=\sum_{i=1}^{m}\alpha_i q_i(\omega)\leq\delta_{m+1}q_{m+1}(\omega).$$

Hence, for almost every $\omega\in\Omega$, $z(\omega)$ is a feasible solution of $W'z\leq q(\omega)$, since $\delta_{m+1}>0$. Now the desired result follows from Th. 15. □

However, the condition given in Cor. 17 is not sufficient in general for the finiteness of $Q(x)$ if W has more than $m+1$ columns, as is shown by the following example:

$$W=(W_1,W_2,W_3,W_4)=\begin{pmatrix}1 & 1 & -1 & -1\\ -1 & 1 & 2 & -2\end{pmatrix}.$$

W is a complete recourse matrix, since W_1 and W_2 are linearly independent and

$$W_3 + W_4 = -W_1 - W_2,$$

and hence

$$\alpha_1 = \alpha_2 = -1$$

$$\delta_3 = \delta_4 = 1.$$

Let $q(\omega) \equiv q$, given by $q_1 = q_2 = q_3 = 1$, $q_4 = -2$, which satisfies

$$\alpha_1 q_1 + \alpha_2 q_2 = -2 \leq \delta_3 q_3 + \delta_4 q_4 = -1.$$

Here

$$W'z \leq q$$

is equivalent to

$$\begin{aligned} z_1 - z_2 &\leq 1 \\ z_1 + z_2 &\leq 1 \\ -z_1 + 2z_2 &\leq 1 \\ -z_1 - 2z_2 &\leq -2. \end{aligned}$$

Summing up the last two inequalities yields

$$z_1 \geq \frac{1}{2};$$

if we add twice the second inequality and the fourth inequality, we get

$$z_1 \leq 0.$$

Hence $\{z \mid W'z \leq q\} = \emptyset$, which implies, by Cor. 16, that $Q(x) = -\infty$.

4. Simple Recourse

Simple recourse is a special case of complete fixed recourse in the following sense:

Definition. $W = (I, -I)$, where I is the $(m \times m)$ identity matrix, is called the *simple recourse matrix*.

This definition says that in the simple recourse model the violations of the original constraints, which may occur after having chosen a decision $x \in X$ and observed the realization of $A(\omega), b(\omega)$, are simply weighed by $q_i(\omega)$. For the simple recourse model it is convenient to write the second stage program as follows:

$$\begin{aligned} Q(x,\omega) = \inf\, & [q^{+\prime}(\omega) y^{+} + q^{-\prime}(\omega) y^{-}] \\ \text{subject to}\quad & y^{+} - y^{-} = b(\omega) - A(\omega)x \\ & y^{+} \geq 0 \\ & y^{-} \geq 0; \quad y^{+}, \quad y^{-} \in \mathbb{R}^m. \end{aligned}$$

Corollary 18. *Given simple recourse and one of the integrability conditions of Th.* 15, *then $Q(x)$ is finite if and only if $q^{+}(\omega) + q^{-}(\omega) \geq 0$ with probability* 1.

Proof. By Th. 15 $Q(x)$ is finite if and only if $\{z \mid W'z \leq q(\omega)\} \neq \emptyset$ with probability 1, i.e. if and only if $\{z \mid -q^-(\omega) \leq z \leq q^+(\omega)\} \neq \emptyset$ with probability 1. This yields the proposition of the Cor. □

The simple recourse model has been studied for various applications all of which have in common that they can be understood as production or allocation problems where only the demand is stochastic. In this case it turns out that we get $Q(x)$, or some equivalent, in a rather explicit form which allows more insight into the structure of the problem than convexity and differentiability do.
Hence we assume that

$$q^+(\omega) \equiv q^+, \quad q^-(\omega) \equiv q^- \quad \text{and} \quad A(\omega) \equiv A;$$

i.e. only $b(\omega)$ is random. According to Cor. 18 we assume that

$$\tilde{q} = q^+ + q^- \geq 0.$$

The following results are due to R. Wets [16]:

Theorem 19. *$Q(x,\omega)$ may be represented as a separable function in*

$$\chi = Ax, \quad \text{i.e.} \quad Q(x,\omega) = \sum_{i=1}^{m} Q_i(\chi_i, \omega).$$

Proof. $Q(x,\omega) = \min\{q^{+\prime}y^+ + q^{-\prime}y^- \mid y^+ - y^- = b(\omega) - Ax, y^+ \geq 0, y^- \geq 0\}$.
By the duality theorem $Q(x,\omega) = \max\{(b(\omega) - Ax)'u \mid -q^- \leq u \leq q^+\}$.
For this program we can immediately find an optimal solution $u^*(\omega)$ with the components

$$u_i^*(\omega) = \begin{cases} q_i^+, & \text{if} \quad (b(\omega) - Ax)_i > 0 \\ -q_i^-, & \text{if} \quad (b(\omega) - Ax)_i \leq 0 \end{cases}$$

$$= \begin{cases} q_i^+, & \text{if} \quad \chi_i < b_i(\omega) \\ -q_i^-, & \text{if} \quad \chi_i \geq b_i(\omega). \end{cases}$$

If we define

$$Q_i(\chi_i, \omega) = \begin{cases} (b_i(\omega) - \chi_i)\, q_i^+, & \text{if} \quad \chi_i < b_i(\omega) \\ -(b_i(\omega) - \chi_i)\, q_i^-, & \text{if} \quad \chi_i \geq b_i(\omega) \end{cases}$$

we get the theorem. □

Theorem 20. *Provided $b(\omega)$ is integrable, $Q(x)$ may be represented as a convex separable function in $\chi = Ax$, i.e.*

$$Q(x) = \sum_{i=1}^{m} Q_i(\chi_i), \quad \textit{where} \quad Q_i(\chi_i) \quad \textit{is convex}, \quad i = 1, \ldots, m.$$

Proof. According to Th. 19 the separability in $\chi = Ax$ follows from

$$Q(x) = \int Q(x,\omega)\,\mathrm{d}P_\omega = \sum_{i=1}^{m} \int Q_i(\chi_i, \omega)\,\mathrm{d}P_\omega$$

i.e.

$$Q_i(\chi_i) = \int Q_i(\chi_i, \omega)\,\mathrm{d}P_\omega.$$

Using the definitions $\tilde{q}=q^+ + q^-$ and $\bar{b}_i=\int b_i(\omega)\mathrm{d}P_\omega$, Th. 19 yields

$$Q_i(\chi_i)=q_i^+ \int_{b_i(\omega)>\chi_i} (b_i(\omega)-\chi_i)\mathrm{d}P_\omega - q_i^- \int_{b_i(\omega)\le\chi_i} (b_i(\omega)-\chi_i)\mathrm{d}P_\omega$$

$$=q_i^+\bar{b}_i - q_i^+\chi_i - \tilde{q}_i \int_{b_i(\omega)\le\chi_i} (b_i(\omega)-\chi_i)\mathrm{d}P_\omega.$$

To prove the convexity of $Q_i(\chi_i)$ it suffices to show that

$$\tilde{q}_i \int_{b_i(\omega)\le\chi_i} (\chi_i - b_i(\omega))\mathrm{d}P_\omega$$

is convex in χ_i.
Since $\tilde{q}_i \ge 0$, we have only to investigate the integral.
Suppose that $\chi_i^1 < \chi_i^2$, $0<\lambda<1$ and $\chi_i^* = \lambda\chi_i^1 + (1-\lambda)\chi_i^2$.
Then

$$\int_{b_i(\omega)\le\chi_i^*} (\chi_i^* - b_i(\omega))\mathrm{d}P_\omega = \lambda \int_{b_i(\omega)\le\chi_i^*} (\chi_i^1 - b_i(\omega))\mathrm{d}P_\omega + (1-\lambda) \int_{b_i(\omega)\le\chi_i^*} (\chi_i^2 - b_i(\omega))\mathrm{d}P_\omega$$

$$=\lambda \int_{b_i(\omega)\le\chi_i^1} (\chi_i^1 - b_i(\omega))\mathrm{d}P_\omega + (1-\lambda) \int_{b_i(\omega)\le\chi_i^2} (\chi_i^2 - b_i(\omega))\mathrm{d}P_\omega$$

$$+\lambda \int_{\chi_i^1<b_i(\omega)\le\chi_i^*} (\chi_i^1 - b_i(\omega))\mathrm{d}P_\omega - (1-\lambda) \int_{\chi_i^*<b_i(\omega)\le\chi_i^2} (\chi_i^2 - b_i(\omega))\mathrm{d}P_\omega$$

$$\le\lambda \int_{b_i(\omega)\le\chi_i^1} (\chi_i^1 - b_i(\omega))\mathrm{d}P_\omega + (1-\lambda) \int_{b_i(\omega)\le\chi_i^2} (\chi_i^2 - b_i(\omega))\mathrm{d}P_\omega,$$

since obviously

$$\int_{\chi_i^1<b_i(\omega)\le\chi_i^*} (\chi_i^1 - b_i(\omega))\mathrm{d}P_\omega \le 0$$

and

$$\int_{\chi_i^*<b_i(\omega)\le\chi_i^2} (\chi_i^2 - b_i(\omega))\mathrm{d}P_\omega \ge 0.$$

Hence

$$\int_{b_i(\omega)\le\chi_i} (\chi_i - b_i(\omega))\mathrm{d}P_\omega$$

is convex in χ_i. □

Suppose now, that there exist α_i, β_i such that

$$\alpha_i \le b_i(\omega) \le \beta_i \quad \text{for all} \quad \omega\in\Omega.$$

Then from

$$Q_i(\chi_i)=q_i^+\bar{b}_i - q_i^+\chi_i - \tilde{q}_i \int_{b_i(\omega)\le\chi_i} (b_i(\omega)-\chi_i)\mathrm{d}P_\omega$$

we know that

$$Q_i(\chi_i)=q_i^+\bar{b}_i - q_i^+\chi_i \quad \text{for} \quad \chi_i<\alpha_i$$

and

$$Q_i(\chi_i)= -q_i^-\bar{b}_i + q_i^-\chi_i \quad \text{for} \quad \chi_i \ge \beta_i,$$

i.e. only on (α_i, β_i) the function $Q_i(\chi_i)$ may be nonlinear.

Thus, it seems desirable to separate the nonlinear and linear terms by constructing a new objective function which yields the same solution set as

$$\sum_{i=1}^{m} Q_i(\chi_i).$$

This may be done by introducing the variables $\chi_{i1}, \chi_{i2}, \chi_{i3}$ and the following constraints:

$$\begin{aligned}
&-\chi_{i1}+\chi_{i2}+\chi_{i3}=\chi_i-\bar{b}_i\\
&\chi_{i1} \qquad\qquad \geq \bar{b}_i-\alpha_i\\
&\chi_{i2} \qquad\qquad \leq \beta_i-\alpha_i\\
&\chi_{i2}\geq 0\\
&\chi_{i3}\geq 0\\
&(\chi_{i1}\geq 0 \quad \text{follows from} \quad \bar{b}_i\geq\alpha_i).
\end{aligned}$$

Let $\mathfrak{B}_i(\chi_i)$ be the set of all feasible $(\chi_{i1}, \chi_{i2}, \chi_{i3})$.
If

$$\psi_i(\chi_i)=\int_{b_i(\omega)\leq\chi_i}(\chi_i-b_i(\omega))\,\mathrm{d}P_\omega$$

and

$$\Phi_i(\chi_{i1},\chi_{i2},\chi_{i3})=\chi_{i3}+\int_{b_i(\omega)\leq\alpha_i+\chi_{i2}}(\chi_{i2}+\alpha_i-b_i(\omega))\,\mathrm{d}P_\omega,$$

then we can state the following

Theorem 21. $\psi_i(\chi_i)=\min\limits_{\mathfrak{B}_i(\chi_i)}\Phi_i(\chi_{i1},\chi_{i2},\chi_{i3})$.

Proof. Let $(\chi_{i1},\chi_{i2},\chi_{i3})\in\mathfrak{B}_i(\chi_i)$ be arbitrarily chosen.

$$\begin{aligned}
\psi_i(\chi_i)&=\int_{b_i(\omega)\leq\chi_i}(\chi_i-b_i(\omega))\,\mathrm{d}P_\omega\\
&=\int_{b_i(\omega)\leq\chi_i}(\bar{b}_i-\chi_{i1}+\chi_{i2}+\chi_{i3}-b_i(\omega))\,\mathrm{d}P_\omega\\
&=\chi_{i3}\cdot P_\omega(b_i(\omega)\leq\chi_i)+(\bar{b}_i-\chi_{i1}-\alpha_i)P_\omega(b_i(\omega)\leq\chi_i)+\\
&\quad+\int_{b_i(\omega)\leq\chi_i}(\chi_{i2}+\alpha_i-b_i(\omega))\,\mathrm{d}P_\omega.
\end{aligned}$$

Since $\chi_{i1}\geq\bar{b}_i-\alpha_i, \chi_{i3}\geq 0$ and $0\leq P_\omega(b_i(\omega)\leq\chi_i)\leq 1$, it follows immediately that

$$\begin{aligned}
\psi_i(\chi_i)&\leq\chi_{i3}+\int_{b_i(\omega)\leq\chi_i}(\chi_{i2}+\alpha_i-b_i(\omega))\,\mathrm{d}P_\omega\\
&\leq\chi_{i3}+\int_{b_i(\omega)\leq\chi_{i2}+\alpha_i}(\chi_{i2}+\alpha_i-b_i(\omega))\,\mathrm{d}P_\omega\\
&=\Phi_i(\chi_{i1},\chi_{i2},\chi_{i3})
\end{aligned}$$

The last inequality is obvious, if $\chi_{i2}+\alpha_i\geq\chi_i$, and if $\chi_{i2}+\alpha_i<\chi_i$, it follows from $\chi_{i2}+\alpha_i-b_i(\omega)<0$ for $b_i(\omega)$ such that $\chi_{i2}+\alpha_i<b_i(\omega)\leq\chi_i$.
Now it suffices to show that there exists $(\chi_{i1}^*,\chi_{i2}^*,\chi_{i3}^*)\in\mathfrak{B}_i(\chi_i)$ such that

$$\Phi_i(\chi_{i1}^*,\chi_{i2}^*,\chi_{i3}^*)=\psi_i(\chi_i).$$

This may be achieved by determining $(\chi_{i1}^*, \chi_{i2}^*, \chi_{i3}^*) \in \mathfrak{B}_i(\chi_i)$, as follows:
For $\chi_i \leq \alpha_i : \chi_{i1}^* = \bar{b}_i - \chi_i, \quad \chi_{i2}^* = \chi_{i3}^* = 0$

$$\Rightarrow \Phi_i(\chi_{i1}^*, \chi_{i2}^*, \chi_{i3}^*) = 0 = \psi_i(\chi_i).$$

For $\alpha_i < \chi_i \leq \beta_i : \chi_{i1}^* = \bar{b}_i - \alpha_i, \quad \chi_{i2}^* = \chi_i - \alpha_i, \quad \chi_{i3}^* = 0$

$$\Rightarrow \Phi_i(\chi_{i1}^*, \chi_{i2}^*, \chi_{i3}^*) = \int_{b_i(\omega) \leq \chi_i} (\chi_i - b_i(\omega))\, dP_\omega = \psi_i(\chi_i).$$

For $\chi_i > \beta_i : \chi_{i1}^* = \bar{b}_i - \alpha_i, \quad \chi_{i2}^* = \beta_i - \alpha_i, \quad \chi_{i3}^* = \chi_i - \beta_i$

$$\Rightarrow \Phi_i(\chi_{i1}^*, \chi_{i2}^*, \chi_{i3}^*) = \chi_i - \beta_i + \beta_i - \bar{b}_i$$
$$= \chi_i - \bar{b}_i = \psi_i(\chi). \quad \square$$

Where $b_i(\omega)$ is integrable and bounded below by α_i, but not essentially bounded above we have

Corollary 22. *Let*

$$\Phi_i(\chi_{i1}, \chi_{i2}) = \int_{b_i(\omega) \leq \alpha_i + \chi_{i2}} (\chi_{i2} + \alpha_i - b_i(\omega))\, dP_\omega$$

and

$$\mathfrak{B}_i(\chi_i) = \{(\chi_{i1}, \chi_{i2}) \mid -\chi_{i1} + \chi_{i2} = \chi_i - \bar{b}_i;\ \chi_{i1} \geq \bar{b}_i - \alpha_i;\quad \chi_{i2} \geq 0\}$$

Then $\psi_i(\chi_i) = \min_{\mathfrak{B}_i(\chi_i)} \Phi_i(\chi_{i1}, \chi_{i2})$.

Proof. The proof follows immediately from that of Th. 21 by setting $\chi_{i3} = 0$ and $\beta_i = +\infty$. $\square$

Now the problem

$$\operatorname*{Min}_{x \in X} \{Q(x) + \bar{c}'x\},$$

where X is usually some convex polyhedral set, may be rewritten as

$$\operatorname{Min} \left\{ \sum_{i=1}^{m} Q_i(\chi_i) + \bar{c}'x \right\}$$
$$\text{subject to} \quad \chi - Ax = 0$$
$$x \in X$$

which is, by the proof of Th. 20, the same as

$$\operatorname{Min} \left\{ \sum_{i=1}^{m} \{q_i^+ \bar{b}_i - q_i^+ \chi_i + \tilde{q}_i \int_{b_i(\omega) \leq \chi_i} (\chi_i - b_i(\omega))\, dP_\omega\} + \bar{c}'x \right\}$$

subject to

$$\chi - Ax = 0$$
$$x \in X.$$

Since, by assumption, $\tilde{q}_i = q_i^+ + q_i^- \geq 0$, $i = 1, \ldots, m$, it follows from Th. 21 that this problem has the same solution set with respect to x as the following one:

$$\operatorname{Min}\left\{\sum_{i=1}^{m}\{q_i^+\chi_{i1}-q_i^+\chi_{i2}+q_i^-\chi_{i3}+\tilde{q}_i\int_{b_i(\omega)\leq\alpha_i+\chi_{i2}}(\chi_{i2}+\alpha_i-b_i(\omega))\mathrm{d}P_\omega\}+\bar{c}'x\right\}$$

subject to $\bar{b}_i-\chi_{i1}+\chi_{i2}+\chi_{i3}-A_ix=0$ where A_i is the i-th row of A

$$\chi_{i1}\geq\bar{b}_i-\alpha_i$$
$$\chi_{i2}\leq\beta_i-\alpha_i$$
$$\chi_{i2}\geq 0$$
$$\chi_{i3}\geq 0$$
$$x\in X.$$

In case that $\beta_i=+\infty$, we set, as in Cor. 22, $\chi_{i3}=0$ and omit the constraint $\chi_{i2}\leq\beta_i-\alpha_i$.
It seems worthwhile mentioning that this representation of the problem implies that contrary to the general complete recourse case, for the simple recourse model, where q^+, q^- and A are constant, only the probability distribution of every $b_i(\omega)$ has to be known, but not their joint distribution. This also means that it does not matter whether the random variables $b_i(\omega)$ are stochastically independent or not.

To illustrate the above result let us give some examples. First suppose that the random variables $b_i(\omega)$ have finite discrete probability distributions, i.e.

$$b_i(\omega)=b_{il}\quad\text{with probability}\quad p_{il},\quad l=1,\ldots,\kappa_i.$$

where $b_{il}<b_{il+1}$ and $p_{il}>0$, $\sum_{l=1}^{\kappa_i}p_{il}=1$, $b_{i1}=\alpha_i$, $b_{i\kappa_i}=\beta_i$.
Then, if $b_{i\nu}\leq\alpha_i+\chi_{i2}\leq b_{i\nu+1}$ for some ν, $1\leq\nu\leq\kappa_i$,

$$\begin{aligned}\int_{b_i(\omega)\leq\chi_{i2}+\alpha_i}(\chi_{i2}+\alpha_i-b_i(\omega))\mathrm{d}P_\omega&=\sum_{\{l\,|\,b_{il}\leq\chi_{i2}+\alpha_i\}}(\chi_{i2}+\alpha_i-b_{il})\,p_{il}\\&=\sum_{\{l:\,b_{il}\leq\chi_{i2}+\alpha_i\}}(b_{i\nu}-b_{il})p_{il}+\sum_{\{l:\,b_{il}\leq\chi_{i2}+\alpha_i\}}(\chi_{i2}+\alpha_i-b_{i\nu})p_{il}\\&=\sum_{l=1}^{\nu-1}(b_{i\nu}-b_{il})p_{il}+\hat{\chi}_{i2}^{\nu}\cdot F_{i\nu}\end{aligned}$$

where

$$0\leq\hat{\chi}_{i2}^{\nu}=\chi_{i2}+\alpha_i-\mathrm{b}_{i\nu}\leq b_{i\nu+1}-b_{i\nu}$$

and

$$F_{i\nu}=\sum_{l=1}^{\nu}p_{il}.$$

Hence the objective function is linear in χ_{i2} on every interval

$$b_{i\nu}-\alpha_i\leq\chi_{i2}\leq b_{i\nu+1}-\alpha_i.$$

If we choose

$$\hat{\chi}_{i2}^{l}=b_{il+1}-b_{il},\quad 1\leq l\leq\nu-1,$$

$$\hat{\chi}_{i2}^{l}=0,\quad l>\nu$$

then

$$\int_{b_i(\omega)\le \chi_{i2}+\alpha_i} (\chi_{i2}+\alpha_i-b_i(\omega))\,\mathrm{d}P=$$

$$=\sum_{l=1}^{\nu-1}\sum_{k=l}^{\nu-1}(b_{ik+1}-b_{ik})p_{il}+\hat{\chi}_{i2}^{\nu}F_{i\nu}$$

$$=\sum_{l=1}^{\nu-1}\sum_{k=l}^{\nu-1}\hat{\chi}_{i2}^{k}p_{il}+\hat{\chi}_{i2}^{\nu}F_{i\nu}$$

$$=\sum_{l=1}^{\nu}\hat{\chi}_{i2}^{l}F_{il}=\sum_{l=1}^{\varkappa_i}\hat{\chi}_{i2}^{l}F_{il}$$

and

$$\sum_{l=1}^{\nu}\hat{\chi}_{i2}^{l}=\chi_{i2},\quad \hat{\chi}_{i2}^{l}\ge 0.$$

Since $F_{il}<F_{il+1}$, $l=1,\ldots,\varkappa_i-1$, it is obvious that

$$\sum_{l=1}^{\varkappa_i}\hat{\chi}_{i2}^{l}F_{il}=\min\sum_{l=1}^{\varkappa_i}\chi_{i2}^{l}F_{il}$$

$$\text{subject to}\quad \sum_{l=1}^{\varkappa_i}\chi_{i2}^{l}=\chi_{i2},\qquad 0\le\chi_{i2}^{l}\le b_{il+1}-b_{il}.$$

Regarding $\tilde{q}_i\ge 0$, therefore we may solve the linear program (provided that X is convex polyhedral):

$$\operatorname{Min}\left\{\sum_{i=1}^{m}\{q_i^+\chi_{i1}-q_i^+\chi_{i2}+q_i^-\chi_{i3}+\tilde{q}_i\sum_{l=1}^{\varkappa_i}\chi_{i2}^{l}F_{il}\}+\bar{c}'x\right\}$$

$$\begin{aligned}\text{subject to}\quad &\bar{b}_i-\chi_{i1}+\chi_{i2}+\chi_{i3}-A_ix=0\\ &\chi_{i1}\ge\bar{b}_i-\alpha_i\\ &0\le\chi_{i2}\le\beta_i-\alpha_i\\ &\chi_{i3}\ge 0\\ &\sum_{l=1}^{\varkappa_i}\chi_{i2}^{l}-\chi_{i2}=0\\ &0\le\chi_{i2}^{l}\le b_{il+1}-b_{il}\\ &x\in X.\end{aligned}$$

Next, we suppose that the random variables $b_i(\omega)$ are uniformly distributed on $[\alpha_i,\beta_i]$, i.e. the distribution is determined by the density function

$$f_i(\tau)=\begin{cases}\dfrac{1}{\beta_i-\alpha_i}, & \alpha_i\le\tau\le\beta_i \qquad (\alpha_i<\beta_i)\\ 0 & \text{otherwise.}\end{cases}$$

Then we get, since

$$0\le\chi_{i2}\le\beta_i-\alpha_i,$$

$$\int_{b_i(\omega)\le\alpha_i+\chi_{i2}}(\alpha_i+\chi_{i2}-b_i(\omega))\,\mathrm{d}P_\omega=\int_{\alpha_i}^{\alpha_i+\chi_{i2}}(\alpha_i+\chi_{i2}-\tau)\cdot\frac{1}{\beta_i-\alpha_i}\,\mathrm{d}\tau$$

$$=\frac{1}{2(\beta_i-\alpha_i)}\cdot\chi_{i2}^2.$$

Hence — as was first pointed out by Beale [2] — we have to solve the convex quadratic program

$$\mathrm{Min}\left\{\sum_{i=1}^{m}\{q_i^+\chi_{i1}-q_i^+\chi_{i2}+q_i^-\chi_{i3}+\frac{\tilde{q}_i}{2(\beta_i-\alpha_i)}\cdot\chi_{i2}^2\}+\bar{c}'x\right\}$$

$$\begin{aligned}\text{subject to}\quad & \bar{b}_i-\chi_{i1}+\chi_{i2}+\chi_{i3}-A_ix=0\\ & \chi_{i1}\geq\bar{b}_i-\alpha_i=\frac{\beta_i-\alpha_i}{2}\\ & 0\leq\chi_{i2}\leq\beta_i-\alpha_i\\ & \chi_{i3}\geq 0\\ & x\in X.\end{aligned}$$

Finally, let us assume that the random variables $b_i(\omega)$ are exponentially distributed with density functions

$$f_i(\tau)=\begin{cases}\lambda_i e^{-\lambda_i\tau}, & \tau\geq 0\\ 0 & \text{otherwise}\end{cases}$$

where $\lambda_i>0$, i.e. $\alpha_i=0$, $\beta_i=+\infty$.

Then we have

$$\int_{b_i(\omega)\leq\alpha_i+\chi_{i2}}(\chi_{i2}+\alpha_i-b_i(\omega))\mathrm{d}P_\omega=\lambda_i\int_0^{\chi_{i2}}(\chi_{i2}-\tau)e^{-\lambda_i\tau}\mathrm{d}\tau$$

$$=\chi_{i2}+\frac{1}{\lambda_i}(e^{-\lambda_i\chi_{i2}}-1).$$

Hence we get the convex program

$$\mathrm{Min}\left\{\sum_{i=1}^{m}\left\{q_i^+\chi_{i1}+q_i^-\chi_{i2}+\frac{\tilde{q}_i}{\lambda_i}e^{-\lambda_i\chi_{i2}}-\frac{\tilde{q}_i}{\lambda_i}\right\}+\bar{c}'x\right\}$$

$$\begin{aligned}\text{subject to}\quad & \bar{b}_i-\chi_{i1}+\chi_{i2}-A_ix=0\\ & \chi_{i1}\geq\bar{b}_i=\frac{1}{\lambda_i}\\ & \chi_{i2}\geq 0\\ & x\in X.\end{aligned}$$

Using the taylor series

$$e^{-\lambda_i\chi_{i2}}=\sum_{\nu=0}^{\infty}(-1)^\nu\frac{\lambda_i^\nu\chi_{i2}^\nu}{\nu!}$$

the objective function of this program may be written as

$$Q(\chi_{.1},\chi_{.2},x)=\left\{\sum_{i=1}^{m}\left\{q_i^+\chi_{i1}-q_i^+\chi_{i2}+\frac{\tilde{q}_i\lambda_i}{2}\chi_{i2}^2+\tilde{q}_i\sum_{\nu=3}^{\infty}(-1)^\nu\frac{\lambda_i^{\nu-1}\chi_{i2}^\nu}{\nu!}\right\}+\bar{c}'x\right\}$$

which may be approximated by its first and second order terms, i.e. by

$$\hat{Q}(\chi_{.1},\chi_{.2},x)=\left\{\sum_{i=1}^{m}\left\{q_i^+\chi_{i1}-q_i^+\chi_{i2}+\frac{\tilde{q}_i\lambda_i}{2}\chi_{i2}^2\right\}+\bar{c}'x\right\}.$$

It is certainly more convenient to solve the approximating quadratic program instead of the more complicated convex program. But then one should have, at least a posteriori, some information on the accuracy of the approximating solution, i.e. one needs at least an a posteriori error bound.

Rewriting the objective functions

$$Q(\chi_{.1},\chi_{.2},x)=\left\{\sum_{i=1}^{m}\left\{q_i^+\chi_{i1}-q_i^+\chi_{i2}+\tilde{q}_i\chi_{i2}+\frac{\tilde{q}_i}{\lambda_i}(e^{-\lambda_i\chi_{i2}}-1)\right\}+\bar{c}'x\right\}$$

$$\hat{Q}(\chi_{.1},\chi_{.2},x)=\left\{\sum_{i=1}^{m}\left\{q_i^+\chi_{i1}-q_i^+\chi_{i2}+\frac{\tilde{q}_i\lambda_i}{2}\chi_{i2}^2\right\}+\bar{c}'x\right\}$$

we have

$$\begin{aligned}\Delta(\chi_{.2})&=Q(\chi_{.1},\chi_{.2},x)-\hat{Q}(\chi_{.1},\chi_{.2},x)\\&=\sum_{i=1}^{m}\left\{\tilde{q}_i[\chi_{i2}+\frac{1}{\lambda_i}(e^{-\lambda_i\chi_{i2}}-1)-\frac{\lambda_i}{2}\chi_{i2}^2]\right\}\\&=\sum_{i=1}^{m}\Delta_i(\chi_{i2}).\end{aligned}$$

From

$$\begin{aligned}&\Delta_i(0)=0,\\&\Delta_i'(\chi_{i2})=\tilde{q}[1-e^{-\lambda_i\chi_{i2}}-\lambda_i\chi_{i2}]\Rightarrow\Delta_i'(0)=0,\quad\text{and}\\&\Delta_i''(\chi_{i2})=\tilde{q}_i[\lambda_i e^{-\lambda_i\chi_{i2}}-\lambda_i]\le 0\quad\text{for}\quad\chi_{i2}\ge 0(\lambda_i>0,\tilde{q}_i\ge 0),\end{aligned}$$

it follows that $\Delta_i(\chi_{i2})\le 0$ and hence

$$\begin{aligned}&\Delta(\chi_{.2})\le 0\quad\text{for}\quad\chi_{.2}\ge 0,\quad\text{i.e.}\\&Q(\chi_{.1},\chi_{.2},x)\le\hat{Q}(\chi_{.1},\chi_{.2},x).\end{aligned}$$

On the other hand, it follows from

$$\tilde{q}_i[\chi_{i2}+\frac{1}{\lambda_i}(e^{-\lambda_i\chi_{i2}}-1)]\ge 0\quad\text{for}\quad\chi_{i2}\ge 0$$

that

$$Q(\chi_{.1},\chi_{.2},x)\ge\sum_{i=1}^{m}\{q_i^+\chi_{i1}-q_i^+\chi_{i2}\}+\bar{c}'x=-q^{+\prime}Ax+q^{+\prime}\bar{b}+\bar{c}'x=L(x).$$

Therefore, if x^*,x^{**} and $\tilde{x}$ are minimal feasible points with respect to $Q,\hat{Q}$ and L, we know from Th. 21, that

$$L(\tilde{x})\le L(x^*)\le Q(\chi_{.1}^*,\chi_{.2}^*,x^*)\le Q(\chi_{.1}^{**},\chi_{.2}^{**},x^{**})=$$

$$=\sum_{i=1}^{m}\{q_i^+\bar{b}_i-q_i^+A_ix^{**}\}+\sum_{i:A_ix^{**}>0}\tilde{q}_i[A_ix^{**}+\frac{1}{\lambda_i}(e^{-\lambda_iA_ix^{**}}-1)]+\bar{c}'x^{**}\le$$

$$\le\hat{Q}(\chi_{.1}^{**},\chi_{.2}^{**},x^{**}).$$

It is obvious that the bounds $Q(\chi_{.1}^{**},\chi_{.2}^{**},x^{**})$ and $L(\tilde{x})$, which are determined by solving a quadratic and a linear program, depend essentially on the data q^+, q^-, $\bar{b}$, A and the feasible set X.

5. Computational Remarks

From the theory developed so far it seems rather difficult to get a numerical solution of a general two-stage program with some arbitrary given joint probability distribution. Take for example a complete fixed recourse problem, the distribution of which is given by a density function. In this case we have to minimize a continuous differentiable convex objective function $Q(x)$ subject to $x \in X$. If X is a bounded convex polyhedral set, this problem can be theoretically solved by the following special method of feasible directions:

Given $x^k \in X$, solve the linear program

$$\operatorname{Min} x' \nabla Q(x^k) \quad \text{subject to} \quad x \in X.$$

If x^k solves this linear program, then x^k solves the original problem $\operatorname{Min} Q(x)$ subject to $x \in X$. Otherwise let y^k be a solution of the linear program. Then solve the one dimensional problem

$$\operatorname{Min} Q(\lambda x^k + (1-\lambda) y^k) \quad \text{subject to} \quad 0 \leq \lambda \leq 1,$$

yielding λ^k. Now restart the procedure with

$$x^{k+1} = \lambda_k x^k + (1-\lambda_k) y^k.$$

It is well known that this method converges to a solution of the original problem $\operatorname{Min} Q(x)$ subject to $x \in X$. However, this procedure involves the repeated evaluation of $Q(x)$ and $\nabla Q(x)$, which as we know from the proof of Th. 12, are given by sums of multiple integrals over sets $\mathfrak{B}_i(x)$, which are polyhedral and depend on x. This type of numerical integration seems not to be completely investigated in numerical analysis; one can only be sure that the amount of work evaluating these integrals is tremendous. Therefore, it does not seem to be reasonable to apply the above procedure. For an alternative approach we may get hints from the examples in section 4. There we have seen that in the simple recourse case, where only b is random, a finite discrete distribution of b leads to a linear program and a uniform distribution of b_i's yields a quadratic program. Finally we gave an a posteriori error estimate for approximating the nonlinear program resulting from exponential distributions by a special quadratic program. From these examples it seems obvious to try the following approach: approximate the given two-stage problem by a special optimization problem which may be handled more easily, e.g. by a linear or quadratic program. Then the only problem consists in finding reasonable error estimates.

Suppose for example that the given two-stage problem is of the simple recourse type where only b is random and the finite distribution of b_i is given by the distribution function $F_i(\tau)$ $(F_i(\alpha_i)=0, F_i(\beta_i)=1)$. According to the last section the objective function of the problem is

$$Q(\chi_{.1}, \chi_{.2}, \chi_{.3}, x) =$$

$$\sum_{i=1}^{m} \{q_i^+ \chi_{i1} - q_i^+ \chi_{i2} + q_i^- \chi_{i3} + \tilde{q}_i \int_{\alpha_i}^{\alpha_i + \chi_{i2}} (\chi_{i2} + \alpha_i - \tau) \mathrm{d}F_i(\tau)\} + \bar{c}' x$$

where $0 \leq \chi_{i2} \leq \beta_i - \alpha_i$. Replacing $F_i(\tau)$ by the discrete distribution

$$\tilde{F}_i(\tau) = F_i(\tau_\nu) \quad \text{for} \quad \tau_\nu \leq \tau < \tau_{\nu+1}, \quad \text{where}$$

$\tau_\nu = \alpha_i + \frac{\nu}{n}(\beta_i - \alpha_i)$, $\nu = 0, 1, \ldots, n$ and n is an arbitrary positive integer, yields a new objective $\tilde{Q}(\chi_{i1}, \chi_{i2}, \chi_{i3}, x)$ which is piecewise linear in χ_{i2} and, as we know, may be replaced by a linear objective function with $2m + m \times n$ instead of $3m$ χ-variables. To get an error estimate for the optimal value of this approximating linear program, we need a bound for

$$|\tilde{Q}(\chi_{i1}, \chi_{i2}, \chi_{i3}, x) - Q(\chi_{i1}, \chi_{i2}, \chi_{i3}, x)|.$$

From the definition of Riemann-Stieltjes integrals we know that

$$\begin{aligned} s_n &= \sum_{\nu=0}^{K} (\chi_{i2} + \alpha_i - \tau_{\nu+1}) [F_i(\tau_{\nu+1}) - F_i(\tau_\nu)] \\ &\leq \int_{\alpha_i}^{\alpha_i + \chi_{i2}} (\chi_{i2} + \alpha_i - \tau) dF_i(\tau) \\ &\leq \sum_{\nu=0}^{K} (\chi_{i2} + \alpha_i - \tau_\nu) [F_i(\tau_{\nu+1}) - F_i(\tau_\nu)] = S_n \end{aligned}$$

where $K \leq n$ is the greatest integer such that

$$\tau_K \leq \alpha_i + \chi_{i2} < \tau_{K+1}.$$

At the same time

$$s_n = \int_{\alpha_i}^{\alpha_i + \chi_{i2}} (\alpha_i + \chi_{i2} - \tau) d\tilde{F}_i(\tau).$$

Therefore, from

$$\left| \int_{\alpha_i}^{\alpha_i + \chi_{i2}} (\alpha_i + \chi_{i2} - \tau) [d\tilde{F}_i(\tau) - dF_i(\tau)] \right| \leq$$

$$\left| S_n - s_n \right| = \left| \sum_{\nu=0}^{K} (\tau_{\nu+1} - \tau_\nu) [F_i(\tau_{\nu+1}) - F_i(\tau_\nu)] \right| \leq \frac{1}{n}(\beta_i - \alpha_i)$$

it follows that

$$|\tilde{Q}(\chi_{i1}, \chi_{i2}, \chi_{i3}, x) - Q(\chi_{i1}, \chi_{i2}, \chi_{i3}, x)| \leq \frac{1}{n} \sum_{i=1}^{m} \tilde{q}_i(\beta_i - \alpha_i);$$

this is the desired error estimate which obvious also remains valid for the optimal values of $\tilde{Q}$ and Q.

If, in the same simple recourse model, $F_i(\tau)$ has a continuous density $f_i(\tau)$, we may try another approximation by replacing $f_i(\tau)$ by a piecewise constant density function $\tilde{f}_i(\tau)$ such that

$$|\tilde{f}_i(\tau) - f(\tau)| \leq \varepsilon \quad \forall \tau \in [\alpha_i, \beta_i].$$

Then

$$\int_{\alpha_i}^{\alpha_i + \chi_{i2}} (\chi_{i2} + \alpha_i - \tau) |\tilde{f}_i(\tau) - f(\tau)| d\tau \leq \frac{(\beta_i - \alpha_i)^2}{2} \varepsilon,$$

and hence

$$|\tilde{Q}-Q| \le \varepsilon \sum_{i=1}^{m} \tilde{q}_i \frac{(\beta_i-\alpha_i)^2}{2} .$$

From the last section we know that for constant densities $f_i(\tau)$ we get quadratic programs. It is now obvious that piecewise constant densities again yield quadratic programs.

It is also evident for the general two stage problem that a finite discrete joint probability distribution yields a linear program. Suppose that we have the general two stage problem

$$\min_{x\in X} \{\bar{c}'x+Q(x)\}$$

where

$$Q(x)=\int Q(x,\omega)\,\mathrm{d}P_\omega, \qquad \bar{c}=\int c(\omega)\,\mathrm{d}P_\omega$$

and

$$Q(x,\omega)=\min\{q'(\omega)y \mid W(\omega)y=b(\omega)-A(\omega)x, y\ge 0\}.$$

Suppose furthermore that P_ω is a finite discrete probability distribution, where the elements $\omega_i\in\Omega, i=1,\ldots,r$, have the probability p_i $(p_i\ge 0; \sum_{i=1}^{r} p_i=1)$. Then it is easily seen, that the two stage problem $\min_{x\in X}\{\bar{c}'x+Q(x)\}$ may be rewritten as

$$\min\{\bar{c}'x+\sum_{i=1}^{r} p_i q'(\omega_i)y^i\}$$

subject to

$$\left.\begin{array}{c} A(\omega_i)x+W(\omega_i)y^i=b(\omega_i) \\ x\in X, \qquad y^i\ge 0 \end{array}\right\} \begin{array}{l} i=1,\ldots,r \\ ; \end{array}$$

which is a linear program if X is convex polyhedral. This linear program has (dual) decomposition structure, where the blocks $W(\omega_i)$ remain unchanged in case of fixed recourse. Therefore, it seems reasonable — from the computational point of view — to approximate any probability distribution by a discrete one. We may conclude from the stability theorems of Kosmol [19] that, under appropriate assumptions on the choice of the discrete probability measures, the optimal values of the resulting linear programs converge to the optimal value of the original problem — at least for compact X and complete fixed recourse.

To get error estimates, let us state the assumptions

A.1) $\{z \mid \exists y\in\mathbb{R}^n: y\ge 0, Wy=z\}=\mathbb{R}^m$.

A.2) $\forall\omega\in\Omega: \{u \mid u\in\mathbb{R}^m, W'u\leqq q(\omega)\}\ne\emptyset$.

A.3) The elements of $A(\omega), b(\omega), q(\omega)$ are square integrable with respect to P_ω.

Hence we require complete fixed recourse so that $Q(x,\omega)$ is finite on Ω and integrable for every $x\in X$ (bounded convex polyhedral). If we define the convex polyhedral cone K_w by

$$K_w=\{q \mid \exists u\in\mathbb{R}^m: W'u-q\le 0\},$$

then A.2) requires that $q(\omega)\in K_w \;\; \forall\omega\in\Omega$.

Let $A_\nu(\omega)$, $b_\nu(\omega)$, $q_\nu(\omega)$ be arrays of the same dimension as $A(\omega)$, $b(\omega)$, $q(\omega)$, but with simple functions as elements. The corresponding objective functions let be $Q_\nu(x,\omega)$ and $Q_\nu(x)=\int Q_\nu(x,\omega)\,\mathrm{d}P_\omega$. Obviously the determination of the simple functions defines a discrete distribution on Ω. We must require, that A.2) is also satisfied for $q_\nu(\omega)$ (at least almost surely). For this purpose, we have to be careful. If for example

$$W=(1,-1)$$

and $q(\omega)$ has the range $R(q)=\{(\xi,\eta)\mid \xi\geqq -2.5;\eta\geqq 2.5\}$, then A.1) and A.2) are satisfied for the original problem. Now let $M=\{(\xi,\eta)\mid -4<\xi\leqq -2, 1\leqq\eta<3\}$ be an interval of some partition. Then $q^{-1}[M]\neq\emptyset$, such that M could have a positive probability. Choosing on M the norm minimal vertex $v=\{(\xi,\eta)\mid\xi=-2, \eta=1\}$ as value of $q_\nu(\omega)$ does not satisfy A.2), since $W'u\leqq q_\nu(\omega)$ yields $-1\leqq u\leqq -2$. But if we choose the norm minimal element of the intersection of M and

$$\begin{aligned}K_w&=\{(\xi,\eta)\mid \exists u: -\eta\leqq u\leqq \xi\}\\ &=\{(\xi,\eta)\mid \xi+\eta\geqq 0\},\end{aligned}$$

i.e. $q_\nu(\omega)=(-2,+2)$, then A.2) is satisfied. In general, the analoguous way (choosing the norm minimal element of the intersection of every interval and K_w) yields a sequence satisfying A.2) too.

Let therefore $(A_\nu(\omega), b_\nu(\omega), q_\nu(\omega))$ be an integrable simple function such that A.2) is satisfied. We want to have an error estimate for the objective function and hence for the optimal value of the approximating problem $\min\{c'x+Q_\nu(x)\mid x\in X\}$, which depends on the approximation of $(A(\omega), b(\omega), q(\omega))$ by $(A_\nu(\omega), b_\nu(\omega), q_\nu(\omega))$, measured by the (generalized) L_2-norm. For any vector-valued function

$$g:\Omega\to\mathbb{R}^k$$

we define

$$\varrho(g)=\sqrt{\int_\Omega \|g(\omega)\|^2\mathrm{d}P_\omega},$$

where $\|\ldots\|$ is the Euclidean norm on $\mathbb{R}^k$. In this connection $\varrho(A)$ means that the matrix $A(\omega)$ is handled as an $(m\cdot n)$-vector.

General error. There are constants α,γ,δ_ν such that

$$|Q(x)-Q_\nu(x)|\leqq [\alpha+\gamma\|x\|]\,\varrho(q-q_\nu)+\delta_\nu[\varrho(b-b_\nu)+\|x\|\,\varrho(A-A_\nu)].$$

This may be seen as follows:
For every convex or concave function

$$\varphi:\mathbb{R}^l\to\mathbb{R}^1 \quad \text{we have}$$

$$\begin{aligned}|\varphi(x)-\varphi(y)|&\leqq \mathrm{Max}\,[|(x-y)'\nabla\varphi(x)|;|(x-y)'\nabla\varphi(y)|]\\ &\leqq \mathrm{Max}\,[\|\nabla\varphi(x)\|\cdot\|x-y\|;\|\nabla\varphi(y)\|\cdot\|x-y\|],\end{aligned}$$

where $\nabla\varphi$ is the gradient (or some subgradient) of φ. Using basic solutions, from the former results we get

$$\begin{aligned}
|\Delta Q_\nu(x,\omega)| &= |Q(x, A(\omega), b(\omega), q(\omega)) - Q(x, A_\nu(\omega), b_\nu(\omega), q_\nu(\omega))| \\
&\leqq |Q(x, A(\omega), b(\omega), q(\omega)) - Q(x, A(\omega), b(\omega), q_\nu(\omega))| \\
&\quad + |Q(x, A(\omega), b(\omega), q_\nu(\omega)) - Q(x, A_\nu(\omega), b_\nu(\omega), q_\nu(\omega))| \\
&= |\tilde{q}_i'(\omega) B_i^{-1}[b(\omega) - A(\omega)x] - \tilde{q}_{\nu j}'(\omega) B_j^{-1}[b(\omega) - A(\omega)x]| \\
&\quad + |\tilde{q}_{\nu j}'(\omega) B_j^{-1}[b(\omega) - A(\omega)x] - \tilde{q}_{\nu k}'(\omega) B_k^{-1}[b_\nu(\omega) - A_\nu(\omega)x]| \\
&\leqq \operatorname*{Max}_{i\in J_1} \|B_i^{-1}[b(\omega) - A(\omega)x]\| \cdot \|q(\omega) - q_\nu(\omega)\| \\
&\quad + \operatorname*{Max}_{i\in J_2} \|B_i^{-1\prime}\tilde{q}_{\nu i}(\omega)\| \cdot \|b(\omega) - b_\nu(\omega) - [A(\omega) - A_\nu(\omega)]x\|,
\end{aligned}$$

where $\{B_i | i\in J_1\}$ and $\{B_i | i\in J_2\}$ are those bases out of W, which for some $x\in X$ and $\omega\in\Omega$ are primal feasible and dual feasible respectively.

From Schwarz's inequality follows with

$$z(x,\omega) = \begin{cases} 0 \quad \text{if} \quad b(\omega) - A(\omega)x = 0 \\ \dfrac{b(\omega) - A(\omega)x}{\|b(\omega) - A(\omega)x\|} \quad \text{else} \end{cases}$$

and

$$r_i(\omega) = \begin{cases} 0 \quad \text{if} \quad q_\nu(\omega) = 0 \\ \dfrac{\tilde{q}_{\nu i}(\omega)}{\|q_\nu(\omega)\|} \quad \text{else} \end{cases}$$

$$\begin{aligned}
|Q(x) - Q_\nu(x)| &\leqq \int_\Omega |\Delta Q_\nu(x,\omega)|\,\mathrm{d}P \\
&\leqq \operatorname*{Max}_{\substack{i\in J_1\\ \omega\in\Omega}} \|B_i^{-1}\cdot z(x,\omega)\| \cdot \varrho(b - Ax) \cdot \varrho(q - q_\nu) \\
&\quad + \operatorname*{Max}_{\substack{i\in J_2\\ \omega\in\Omega}} \|B_i^{-1\prime} r_i(\omega)\| \cdot \varrho(q_\nu) \cdot \varrho(b - b_\nu + (A_\nu - A)x)
\end{aligned}$$

Hence, since $\varrho(g+h) \leqq \varrho(g) + \varrho(h)$, for

$$\alpha = \operatorname*{Max}_{\substack{i\in J_1\\ \omega\in\Omega}} \|B_i^{-1} z(x,\omega)\|\, \varrho(b)$$

$$\gamma = \operatorname*{Max}_{\substack{i\in J_1\\ \omega\in\Omega}} \|B_i^{-1} z(x,\omega)\|\, \varrho(A)$$

$$\delta_\nu = \operatorname*{Max}_{\substack{i\in J_2\\ \omega\in\Omega}} \|B_i^{-1\prime} r_i(\omega)\|\, \varrho(q_\nu),$$

it follows that

$$|Q(x) - Q_\nu(x) \leqq [\alpha + \gamma\|x\|]\,\varrho(q - q_\nu) + \delta_\nu\,[\varrho(b - b_\nu) + \|x\|\,\varrho(A - A_\nu)],$$

which was the hypothesis.

It must be mentioned that determining the constants α and γ leads in general to a considerable amount of work, since it implies more or less the inversion of all nonsingular $(m\times m)$-submatrices of W. This difficulty diminishes rapidly for certain special cases. Determining δ_ν (or at least an upper bound) is not dif-

ficult, since $B_i^{-1'}r_i(\omega)$ is a feasible u-part of the set $\{(u,q)|\, W'u \leqq q, \|q\| \leqq 1\}$, which is bounded according to A. 1).
Assume simple recourse, i.e. $W=(I,-I)$. Then

$$|Q(x)-Q_\nu(x)| \leqq [\varrho(\mathrm{b})+\|x\|\,\varrho(A)]\varrho(q-q_\nu)+\varrho(q_\nu)[\varrho(b-b_\nu)+\|x\|\,\varrho(A-A_\nu)].$$

Namely $W=(I,-I)$ implies that for every basis B_i out of W $\|B_i^{-1}z(x,\omega)\| = \|z(x,\omega)\|$ and, by definition, $\|z(x,\omega)\| \leqq 1$. And for every $i \in J_2$, $B_i^{-1'}r_i(\omega)$ is a feasible u-part of $\{(u,q)|\, W'u \leqq q, \|q\| \leqq 1\}$, which for

$$q=\begin{pmatrix} q^+ \\ q^- \end{pmatrix}=\begin{pmatrix} q_1^+ \\ \vdots \\ q_m^+ \\ q_1^- \\ \vdots \\ q_m^- \end{pmatrix} \text{ may be written as}$$

$$\{(u,q^+,q^-)|\, -q^- \leqq u \leqq q^+ ; \|q^+\|^2+\|q^-\|^2 \leqq 1\}.$$

Obviously every feasible u satisfies $\|u\| \leqq 1$. Using these observations, we get for the constants defined in the proof of the general error formula

$$\alpha \leqq \varrho(b)$$
$$\gamma \leqq \varrho(A)$$
$$\delta_\nu \leqq \varrho(q_\nu).$$

For the general complete recourse problem we get out of the difficulties of determining α and γ, if $q(\omega)$ is constant, as is readily seen above. Moreover we can weaken assumption A. 3) to integrability instead of square integrability. Defining a generalized L_1-norm for vector valued functions as

$$\mu(g)=\int_\Omega \|g(\omega)\|\,\mathrm{d}P,$$

we get the error estimate:
For constant $q(\omega)\equiv q$ there is a constant δ such that

$$|Q(x)-Q_\nu(x)| \leqq \delta[\mu(b-b_\nu)+\|x\|\,\mu(A-A_\nu)].$$

From $|\Delta Q_\nu(x,\omega)|$ above we get

$$|\Delta Q(x,\omega)| \leqq \operatorname*{Max}_{i\in J_2} \|B_i^{1'}\tilde{q}_i\| \cdot \|b(\omega)-b_\nu(\omega)-[A(\omega)-A_\nu(\omega)]x\|$$

and therefore

$$|Q(x)-Q_\nu(x)| \leqq \operatorname*{Max}_{i\in J_2} \|B_i^{-1'}\tilde{q}_i\|\,[\mu(b-b_\nu)+\|x\|\,\mu(A-A_\nu)].$$

For $i \in J_2$, $B_i^{-1'}\tilde{q}_i$ is feasible in $\{u|\, W'u \leqq q\}$, which is bounded.

Every one of the error estimates above becomes independent of x, if $A(\omega)\equiv A$ and $q(\omega)\equiv q$ are constant, i.e. in this case we get uniform convergence of $\{Q_\nu(x)\}$ on X, if the simple functions chosen converge to the remaining random variables with respect to the appropriate norm.
One might object that this type of approximation is not practicable, since the size of the approximating problems becomes very large. If for example in (A,b,q) 50 random variables with a joint probability distribution are involved, and if we discretize in such a way that for every random variable 10 realizations occur, then we should have $m\times 10^{50}$ constraints, which cannot be handled. However, in most of the practical problems there is a small number of random variables $t_1,t_2,\ldots,t_r$, where often $r\leqq 5$, and (A,b,q) depend on $t_1,\ldots,t_r$. If this dependence looks like

$$A(t)=A_0+A_1t_1+\cdots+A_rt_r$$
$$b(t)=b_0+b_1t_1+\cdots+b_rt_r$$
$$q(t)=q_0+q_1t_1+\cdots+q_rt_r,$$

then the discretization can be carried through with respect to the random vector t yielding problems of a size which can be handled today.
It is obvious, then, that all above mentioned error estimates can be expressed on $\varrho(t-t_{(\nu)})$ or $\mu(t-t_{(\nu)})$, where $t_{(\nu)}$ is a simple function.

6. Another Approach to Two-Stage Programming

In a more general framework of stochastic input-output systems Marti [11] found some interesting results. The generalization consists essentially in studying stochastic linear programs on arbitrary topological linear spaces instead of Euclidean spaces. Then it turns out to be useful to investigate more thoroughly the optimal value of the second stage problem—which we called $Q(x,\omega)$. Restricted to Euclidean spaces Marti is first concerned with the complete fixed recourse objective function, i.e. his second stage program is

$$Q(x,\omega)=\min\{q'y\mid Wy=b(\omega)-A(\omega)x, y\geq 0\}$$

where W is a complete recourse matrix and $q\geq 0$.
The investigation of the optimal value of the recourse program $m(z)=\min\{q'y\mid Wy=z, y\geq 0\}$ for $z\in\mathbb{R}^m$ yields the following statements:

Lemma 23. *Under the above-mentioned assumption*:
a) $0\leq m(z)<\infty$
b) $m(0)=0$
c) $m(\lambda z)=\lambda m(z)\quad \forall z\in\mathbb{R}^m,\quad \forall\lambda\geq 0.$
d) $m(z_1+z_2)\leq m(z_1)+m(z_2)\quad \forall z_1,z_2\in\mathbb{R}^m.$

Proof. a) Follows immediately from $q\geq 0$ and the complete recourse assumption.
b) $y=0$ yields a solution.

c) Suppose $\lambda > 0$ and let y be a solution corresponding to z; then $m(z) = q'y$. λy is feasible with respect to λz, hence $m(\lambda z) \leq q'(\lambda y) = \lambda m(z)$.
If $m(\lambda z) < \lambda m(z)$, let $\hat{y}$ be a solution with respect to λz, i.e. $m(\lambda z) = q'\hat{y}$. Then $\frac{1}{\lambda}\hat{y}$ is feasible with respect to z and therefore $m(z) \leq \frac{1}{\lambda} q'\hat{y} = \frac{1}{\lambda} m(\lambda z) < m(z)$.
From this contradiction it follows that $m(\lambda z) = \lambda m(z)$.

d) Follows from the convexity theorem (Th. 0.7)

$$m(\lambda z_1 + (1-\lambda) z_2) \leq \lambda m(z_1) + (1-\lambda) m(z_2) \quad \text{and} \quad \text{c)} \quad \text{for } \lambda = \tfrac{1}{2}. \square$$

Definition. Let $K \subset \mathbb{R}^m$ be convex and such that $0 \in \mathbb{R}^m$ is an interior point of K. Then $Q_K(z) = \inf\{\lambda > 0 \mid \frac{z}{\lambda} \in K\}$ is called the *Minkowski functional of* K (on $\mathbb{R}^m$).

Theorem 24. *Under the assumptions of Lemma 23* $K = \{z \mid m(z) \leq 1\}$ *is a convex polyhedral set, which has* $0 \in \mathbb{R}^m$ *as an interior point.*

Proof. From the convexity theorem (Th. 0.7) we know that
1) $m(z)$ is piecewise linear and convex on $\mathbb{R}^m$ and hence
2) continuous on $\mathbb{R}^m$ (Th. II. 12).

Therefore K is convex polyhedral and $\{z \mid m(z) < 1\} \subset K$ is open and contains $0 \in \mathbb{R}^m$ due to Lemma 23 b). $\square$

Theorem 25. $m(z)$ *is the Minkowski functional of the set* K *defined in Th.* 24.

Proof. The Minkowski functional of K is defined as

$$\begin{aligned} Q_K(z) &= \inf\left\{\lambda > 0 \,\middle|\, \frac{z}{\lambda} \in K\right\} \\ &= \inf\left\{\lambda > 0 \,\middle|\, m\left(\frac{z}{\lambda}\right) \leq 1\right\} \quad \text{due to Th. 24} \\ &= \inf\{\lambda > 0 \mid m(z) \leq \lambda\} \quad \text{due to Lemma 23c)} \\ &= m(z). \square \end{aligned}$$

On the other hand we have the converse statement.

Theorem 26. *For every Minkowski functional* $Q_K(z)$ *of a convex polyhedral set* $K \subset \mathbb{R}^m$ *(which contains by definition* $0 \in \mathbb{R}^m$ *as an interior point) there exists a complete recourse matrix* W *and a vector* $q \geq 0$ *such that* $Q_K(z) = m(z) = \min\{q'y \mid Wy = z, y \geq 0\}$.

Proof. The convex polyhedral set K is the direct sum of a convex polyhedron and a convex polyhedral cone, i.e. there are vectors $p_i \in \mathbb{R}^m, i = 1, \ldots, r$, and $k_j \in \mathbb{R}^m$, $j = 1, \ldots, s$, so that

$$K = \left\{v \mid v = \sum_{i=1}^{r} \alpha_i p_i + \sum_{j=1}^{s} \beta_j k_j; \alpha_i \geq 0, \beta_j \geq 0, \sum_{i=1}^{r} \alpha_i = 1\right\}.$$

Since $0 \in \mathbb{R}^m$ is an interior point of K,
for every $z \in \mathbb{R}^m$ there exist $v \in K$ and $\lambda > 0$ such that

$$z = \lambda v = \lambda \left\{\sum_{i=1}^{r} \alpha_i p_i + \sum_{j=1}^{s} \beta_j k_j\right\}, \quad \text{where} \quad \alpha_i \geq 0, \quad \beta_j \geq 0, \quad \sum_{i=1}^{r} \alpha_i = 1.$$

Hence for the matrix

$$W=(p_1,\ldots,p_r,k_1,\ldots,k_s)$$

and

$$y=(\lambda\alpha_1,\ldots,\lambda\alpha_r,\lambda\beta_1,\ldots,\lambda\beta_s)'\geq 0$$

we have

$$Wy=z,$$

which implies that W is a complete recourse matrix.
Furthermore we have

$$\sum_{i=1}^{r} y_i=\lambda\sum_{i=1}^{r}\alpha_i=\lambda.$$

Hence

$$\begin{aligned}\varrho_K(z)&=\inf\left\{\lambda>0\,\middle|\,\frac{z}{\lambda}\in K\right\}\\&=\inf\{\lambda>0\mid \lambda v=z, v\in K\}\\&=\inf\left\{\lambda>0\mid Wy=z, y\geq 0, \lambda=\sum_{i=1}^{r} y_i\right\}\\&=\inf\left\{\sum_{i=1}^{r} y_i\mid Wy=z, y\geq 0\right\},\end{aligned}$$

which proves the theorem. □

According to this correspondence between optimal values of linear programs and Minkowski functionals one can take advantage of the knowledge on Minkowski functionals, which have been investigated in functional analysis, to get continuity and differentiability statements analogous to theorems 10 and 12 in the complete fixed recourse case with $q\geq 0$.
Recently Marti [12] continued this approach requiring instead of $q\geq 0$ the more general condition of Cor. 16, which guarantees the existence of a solution of the second stage program in the complete recourse case. Then the functional

$$\varrho(z)=\min\{q'y\mid Wy=z, y\geq 0\}$$

is, in general, not a Minkowski functional, since $\varrho(z)$ may be negative. But $\varrho(z)$ has still the following properties:

Lemma 27. a) $\varrho(\lambda z)=\lambda\varrho(z)\quad \forall z\in\mathbb{R}^m,\quad \forall\lambda>0$
b) $\varrho(z_1+z_2)\leq\varrho(z_1)+\varrho(z_2)$
c) $\varrho(z)$ *is continuous.*

Proof. a), b) are proved in the same way as in Lemma 23;
c) follows from the complete recourse and finiteness assumptions and the convexity of $\varrho(z)$. □

Lemma 28. *There exist vectors* $g_i\in\mathbb{R}^m$, $i=1,\ldots,r$, *such that* $\varrho(z)=\max_i\{g_i'z\}$.

Proof. According to our assumptions — complete recourse and existence of solution — for every $z \in R^m$ we have a feasible optimal basis B in W, i.e. B is a $m \times m$ nonsingular submatrix of W such that

$$B^{-1}z \geq 0$$

and

$$\varrho(z) = \tilde{q}' B^{-1} z,$$

where $\tilde{q}$ consists of the m components of q belonging to B. Optimality of B means, according to the simplex criterion,

$$q' - \tilde{q}' B^{-1} W \geq 0.$$

Let $B_1, \ldots, B_r$ be all "optimal" bases in W, i.e. all nonsingular $m \times m$ submatrices of W fulfilling

$$q' - \tilde{q}_i' B_i^{-1} W \geq 0.$$

Due to the duality theorem

$$\varrho(z) = \max\{z'g \mid W'g \leq q\}.$$

Since $g_i = (\tilde{q}_i' B_i^{-1})'$, $i = 1, \ldots, r$, is feasible in this dual program, we have $z'g_i \leq \varrho(z)$, $i = 1, \ldots, r$, where equality holds for at least one g_i. □

According to Lemma 28 we may rewrite $Q(x)$ as

$$Q(x) = \int \varrho(b(\omega) - A(\omega)x) \mathrm{d}P_\omega = \int \max_{1 \leq i \leq r} \{g_i'(b(\omega) - A(\omega)x)\} \mathrm{d}P_\omega.$$

From this representation we may conclude an error estimate for the discretization mentioned in Section 5 at least for the case when (A, b) has a finite probability distribution and X is bounded.

Suppose that $\mathfrak{A}$ is an interval in the (A, b)-space with $P(\mathfrak{A}) = 1$, and is partitioned into intervals $\mathfrak{A}_j$—i.e. $\mathfrak{A}_i \cap \mathfrak{A}_j = \emptyset$ and $\bigcup_{j=1}^{s} \mathfrak{A}_j = \mathfrak{A}$ — such that, for $d_1, d_2 \in \mathfrak{A}_j$, $\|d_1 - d_2\| \leq \delta$.

Let $p_j = P_\omega(\{\omega \mid (A(\omega), b(\omega)) \in \mathfrak{A}_j\})$.

For some $x \in X$ let $(A_{jk}, b_{jk}) \in \mathfrak{A}_j$, $k = 1, 2$, be such that

$$\varrho(b_{j1} - A_{j1}x) \leq \varrho(b - Ax) \leq \varrho(b_{j2} - A_{j2}x), \quad \forall (A, b) \in \mathfrak{A}_j.$$

Then,

$$\sum_{j=1}^{s} p_j \varrho(b_{j1} - A_{j1}x) \leq Q(x) = \int \varrho(b(\omega) - A(\omega)x) \mathrm{d}P_\omega \leq \sum_{j=1}^{s} p_j \varrho(b_{j2} - A_{j2}x).$$

If we choose some $(A_j, b_j) \in \mathfrak{A}_j$ we can approximate $Q(x)$ by

$$\tilde{Q}(x) = \sum_{j=1}^{s} p_j \varrho(b_j - A_j x),$$

which yields a linear program, as shown in sec. 5; we then get the error estimate

$$|\tilde{Q}(x)-Q(x)| \leq \sum_{j=1}^{s} p_j[\varrho(b_{j2}-A_{j2}x)-\varrho(b_{j1}-A_{j1}x)]$$
$$\leq \sum_{j=1}^{s} p_j[\max_i \|g_i\| \{\delta+\delta\|x\|\}],$$

using the convexity of ϱ (see Th. 0.11)

$$=\max_i \|g_i\| \{\delta+\delta\|x\|\}$$
$$\leq C\cdot\delta, \quad \text{since} \quad \|x\| \quad \text{is bounded.}$$

This justifies the discretization, because if $\hat{x}$ and $\tilde{x}$ are solutions of

$$\min_{x\in X}\{\bar{c}'x+Q(x)\} \quad \text{and} \quad \min_{x\in X}\{\bar{c}'x+\tilde{Q}(x)\} \quad \text{respectively,}$$

and, without loss of generality, $\bar{c}'\hat{x}+Q(\hat{x})<\bar{c}'\tilde{x}+\tilde{Q}(\tilde{x})$, then

$$0<\bar{c}'\tilde{x}+\tilde{Q}(\tilde{x})-\bar{c}'\hat{x}-Q(\hat{x})\leq\tilde{Q}(\hat{x})-Q(\hat{x})\leq C\cdot\delta.$$

But it must be pointed out that there are still some difficulties in calculating this error bound, since determining

$$\max_i \|g_i\| = \max_i \|\tilde{q}_i' B_i^{-1}\| \leq \|q\|\cdot\max_i \|B_i^{-1}\|$$

is not at all trivial, if one does not want to determine all inverses of bases in W. For the special case where $q>0$ and $\bar{c}'x\geq 0$ on X, the boundedness assumption on X is not restrictive, as is indicated by the following theorem, because there then exists $\tau>0$ such that $\varrho(z)\geq\tau\|z\|$, due to Lemma 28.

Theorem 29. *If $\bar{c}'x\geq 0$ for $x\in X$ and there is a real $\tau>0$ such that $\varrho(z)>\tau\|z\|$ and if*
$P_\omega(\{\omega|A(\omega)x=0\})<1$ for every $x\neq 0$, then there is a compact set $\mathfrak{B}$ such that

$$\inf_X \{\bar{c}'x+Q(x)\}=\inf_{X\cap\mathfrak{B}}\{\bar{c}'x+Q(x)\}.$$

Proof. Let

$$\varphi(x)=\int\|A(\omega)x\|\,\mathrm{d}P_\omega.$$

Then, from the assumption

$$P_\omega(\{\omega|A(\omega)x=0\})<1 \quad \text{for} \quad x\neq 0$$

it follows that

$$\varphi(x)>0 \quad \text{for} \quad x\neq 0 \quad \text{and} \quad \varphi(0)=0,$$
$$\varphi(\lambda x)=|\lambda|\,\varphi(x),$$
$$\varphi(x+y)\leq\varphi(x)+\varphi(y).$$

Hence $\varphi(x)$ is a norm on $\mathbb{R}^n$ and there exists a real $\kappa>0$ such that

$$\varphi(x)\geq\kappa\|x\|.$$

Now for $x \in X$

$$\bar{c}'x + Q(x) \geq Q(x) = \int \varrho(b(\omega) - A(\omega)x)\,dP_\omega \geq \tau \int \|b(\omega) - A(\omega)x\|\,dP_\omega$$
$$\geq \tau \int \big|\,\|b(\omega)\| - \|A(\omega)x\|\,\big|\,dP_\omega \geq \tau \left| \int \|b(\omega)\|\,dP_\omega - \int \|A(\omega)x\|\,dP_\omega \right|$$
$$\geq \tau\left(\varphi(x) - \int \|b(\omega)\|\,dP_\omega\right)$$
$$\geq \tau\kappa\|x\| - \tau E_{P_\omega}\|b\|.$$

For an arbitrary $\hat{x} \in X$ and every x such that

$$\|x\| > \frac{1}{\kappa} E_{P_\omega}\|b\| + \frac{1}{\tau\kappa}\,(Q(\hat{x}) + \bar{c}'\hat{x}) = r(\hat{x}),$$

we have

$$\bar{c}'x + Q(x) \geq \tau\kappa\|x\| - \tau E_{P_\omega}\|b\| > Q(\hat{x}) + \bar{c}'\hat{x}.$$

Hence

$$\inf\{\bar{c}'x + Q(x) \mid x \in X\} = \inf\{\bar{c}'x + Q(x) \mid x \in X, \|x\| \leq r(\hat{x})\}. \quad \square$$

In the complete recourse case — $q \geq 0$ and x bounded is not required — there are upper and lower bounds of $\inf_{x \in X}\{\bar{c}x + Q(x)\}$ given by inequalities first proved by Madansky [9], which are obtained by solving the linear program (provided that X is a convex polyhedral set) resulting from replacing the random variables $A(\omega)$, $b(\omega)$ in the two-stage program by their expectations $\bar{A}$, $\bar{b}$.

Theorem 30. *Suppose that $\hat{x}$ is a solution of the problem*

$$\min_{x \in X}\{\bar{c}'x + \varrho(\bar{b} - \bar{A}x)\}.$$

Then

$$\bar{c}'\hat{x} + \varrho(\bar{b} - \bar{A}\hat{x}) \leq \inf_{x \in X}\{\bar{c}'x + \int \varrho(b(\omega) - A(\omega)x)\,dP_\omega\}$$
$$\leq \bar{c}'\hat{x} + \int \varrho(b(\omega) - A(\omega)\hat{x})\,dP_\omega.$$

Proof. By Lemma 28

$$\varrho(z) = \max_i g_i' z.$$

Choose g_{i_0} so that

$$g_{i_0}'(\bar{b} - \bar{A}x) = \varrho(\bar{b} - \bar{A}x).$$

Then

$$\varrho(\bar{b} - \bar{A}x) = \int g_{i_0}'(b(\omega) - A(\omega)x)\,dP_\omega$$
$$\leq \int \max_i g_i'(b(\omega) - A(\omega)x)\,dP_\omega$$
$$= \int \varrho(b(\omega) - A(\omega)x)\,dP_\omega \qquad \forall x \in X$$

and, therefore,

$$\bar{c}'\hat{x} + \varrho(\bar{b} - \bar{A}\hat{x}) \leq \bar{c}'x + \varrho(\bar{b} - \bar{A}x) \leq \bar{c}'x + \int \varrho(b(\omega) - A(\omega)x)\,dP_\omega \qquad \forall x \in X,$$

which yields the first inequality. The second one is trivial. $\square$

In the special case, where only $b(\omega)$ is random, we get the following inequalities derived from A. Madansky [9]:

Theorem 31. *Under the assumptions of Th.* 30 *and for deterministic matrices A and c the following inequalities hold*:

$$\begin{aligned} c'\hat{x}+\varrho(\bar{b}-A\hat{x}) &\leq \int \inf_{x\in X}\{c'x+\varrho(b(\omega)-Ax)\}\,\mathrm{d}P_\omega \\ &\leq \inf_{x\in X}\{c'x+\int\varrho(b(\omega)-Ax)\,\mathrm{d}P_\omega\} \\ &\leq c'\hat{x}+\int\varrho(b(\omega)-A\hat{x})\,\mathrm{d}P_\omega. \end{aligned}$$

Proof. Observe that under our assumptions

$\varphi(b)=\inf_{x\in X}\{c'x+\varrho(b-Ax)\}$ is a convex function of b,

because

$$\varphi(b)=\inf\{c'x+q'y\,|\,Ax+Wy=b,\,y\geq 0,\,x\in X\}.$$

Therefore

$$\varphi(b)\geq\varphi(\bar{b})+g'(b-\bar{b}),$$

where g is the gradient of φ in $\bar{b}$ (or some subgradient).
Integration of this inequality with respect to P_ω yields

$$\int\varphi(b(\omega))\,\mathrm{d}P_\omega\geq\varphi(\bar{b})$$

which is the first inequality of the theorem, whereas the other ones are trivial. □

As we know from Ch. I

$$\alpha=\int\inf_{x\in X}\{c'x+\varrho(b(\omega)-Ax)\}\,\mathrm{d}P_\omega$$

is the expectation of the optimal value in the "wait and see" case, and

$$\beta=\inf_{x\in X}\{c'x+\int\varrho(b(\omega)-Ax)\,\mathrm{d}P_\omega\}$$

is the optimal value obtained by the two-stage model in the "here and now" situation. M. Avriel and A.C. Williams [1] call the difference $\beta-\alpha$ the expected value of perfect information (EVPI). From Th. 31 we get bounds for EVPI:

$$0\leq\beta-\alpha\leq\int\varrho(b(\omega)-A\hat{x})\,\mathrm{d}P_\omega-\varrho(\bar{b}-A\hat{x})$$

where $\hat{x}$ can be determined by solving a linear program. Unfortunately these bounds for the EVPI are not valid if also $A(\omega)$ and $c(\omega)$ are random, because the first inequality of Th. 31 does not hold in general, as the following example shows:
Example. Let $W=(1,-1)$, $q^+=1$, $q^-=0$

$$X=\{x\,|\,x\in\mathbb{R},\,x\geq 0\}\quad\text{and}$$

$$P_\omega((A,b,c)=(1,2,2))=P_\omega((A,b,c)=(3,12,2))=\tfrac{1}{2}.$$

Then

$$\bar{A}=2, \quad \bar{b}=7, \quad \bar{c}=2 \quad \text{and}$$

$$\bar{c}\hat{x}+\varrho(\bar{b}-\bar{A}\hat{x})=\operatorname*{Min}_{x\in X}\{\bar{c}x+q^+y^+ \mid \bar{A}x+y^+-y^-=\bar{b}, y^+\geq 0, y^-\geq 0\}$$

$$=7, \quad \text{where} \quad \hat{x}=\frac{7}{2}.$$

But

$$\alpha=\int \inf_{x\in X}\{cx+\varrho(b(\omega)-A(\omega)x)\}\,\mathrm{d}P_\omega=$$

$$\frac{1}{2}\operatorname*{Min}_{x\in X}\{2x+y_1^+ \mid x+y_1^+-y_1^-=2, y_1^+\geq 0, y_1^-\geq 0\}+$$

$$+\frac{1}{2}\operatorname*{Min}_{x\in X}\{2x+y_2^+ \mid 3x+y_2^+-y_2^-=12, y_2^+\geq 0, y_2^-\geq 0\}=$$

$$=\frac{1}{2}\cdot 2+\frac{1}{2}\cdot 8=5<\bar{c}\hat{x}+\varrho(\bar{b}-\bar{A}\hat{x}).$$

Hence the first inequality of Th. 31 does not hold in this case.
Further

$$\beta=\inf_{x\in X}\{\bar{c}'x+\int\varrho(b(\omega)-A(\omega)x)\,\mathrm{d}P_\omega\}$$

$$=\operatorname{Min}\Big\{2x+\frac{1}{2}y_1^+ +\frac{1}{2}y_2^+ \,\Big|\, x+y_1^+-y_1^-=2;$$

$$3x+y_2^+-y_2^-=12; x\geq 0, y_i^+\geq 0, y_i^-\geq 0\Big\}$$

$$=7 \quad (\text{choosing} \quad x=2, \quad y_2^+=6) \quad \text{and}$$

$$c'\hat{x}+\int\varrho(b(\omega)-A(\omega)\hat{x})\,\mathrm{d}P_\omega=7+\frac{3}{4}=7{,}75.$$

Hence,

$$\int\varrho(b(\omega)-A(\omega)\hat{x})\,\mathrm{d}P_\omega-\varrho(\bar{b}-\bar{A}\hat{x})=0{,}75<\beta-\alpha=2,$$

which shows, that the bound given above for the EVPI is not valid in the more general case of random $A(\omega)$ and $c(\omega)$.

IV. Chance Constrained Programming

1. Convexity Statements

Whereas two-stage problems, as we have seen in the last chapter, are rather well-behaved from the viewpoint of optimization theory as far as convexity, continuity and differentiability are concerned, this is in general not true for chance constrained programming problems. There are essentially two different versions of chance constrained programs, namely either

$$\begin{aligned} &\min \varphi(x) \\ &\text{subject to } P_\omega(\{\omega \mid A(\omega)x \geq b(\omega)\}) \geq \alpha \\ &x \in X \end{aligned} \tag{1}$$

or

$$\begin{aligned} &\min \varphi(x) \\ &\text{subject to } P_\omega(\{\omega \mid A_i(\omega)x \geq b_i(\omega)\}) \geq \alpha_i, \quad i=1,\dots,m \\ &x \in X \end{aligned} \tag{2}$$

where $A_i(\omega)$ indicates the i-th row of $A(\omega)$ and $b_i(\omega)$ is the i-th component of $b(\omega)$. Given that $\varphi(x)$ is a convex function and X is a convex set, the main question is whether the sets

$$X(\alpha) = \{x \mid P_\omega(\{\omega \mid A(\omega)x \geq b(\omega)\}) \geq \alpha\}$$

and

$$X_i(\alpha_i) = \{x \mid P_\omega(\{\omega \mid A_i(\omega)x \geq b_i(\omega)\}) \geq \alpha_i\} \quad \text{are convex.}$$

The following example shows that the convexity of these sets cannot be guaranteed in general.

Example 1. Let $\begin{pmatrix} a \\ b \end{pmatrix}$ be a two-dimensional random variable with the discrete distribution

$$P\left[\begin{pmatrix} a \\ b \end{pmatrix} = \begin{pmatrix} -3 \\ -1 \end{pmatrix}\right] = \frac{1}{3}; \qquad P\left[\begin{pmatrix} a \\ b \end{pmatrix} = \begin{pmatrix} 3 \\ 2 \end{pmatrix}\right] = \frac{2}{3}.$$

To get

$$P(\{(a,b) \mid ax \geq b\})$$

for a certain value of $x \in \mathbb{R}$, we have to check which of the two constraints

$$-3x \geq -1 \quad \text{and} \quad 3x \geq 2$$

is satisfied. Obviously

$$x \le \frac{1}{3} \quad \text{yields} \quad -3x \ge -1 \quad \text{and} \quad 3x \ngeq 2,$$

$$\frac{1}{3} < x < \frac{2}{3} \quad \text{yields} \quad -3x \ngeq -1 \quad \text{and} \quad 3x \ngeq 2$$

$$\text{and} \quad \frac{2}{3} \le x \quad \text{yields} \quad -3x \ngeq -1 \quad \text{and} \quad 3x \ge 2.$$

Hence

$$P(\{(a,b) \mid ax \ge b\}) = \begin{cases} \frac{1}{3} & \text{for} \quad x \le \frac{1}{3} \\ 0 & \text{for} \quad \frac{1}{3} < x < \frac{2}{3} \\ \frac{2}{3} & \text{for} \quad x \ge \frac{2}{3}, \end{cases}$$

which implies, that

$$X(\alpha) \quad \text{is} \begin{cases} \text{disconnected and hence not convex for} & 0 < \alpha \le \frac{1}{3} \\ \text{convex} & \text{for} \quad \frac{1}{3} < \alpha \le \frac{2}{3} \\ \text{empty} & \text{for} \quad \alpha > \frac{2}{3}. \end{cases}$$

The following theorem is the only general convexity statement on $X(\alpha)$ that can be made disregarding the probability distribution P_ω.

Theorem 1. *$X(0)$ and $X(1)$ (resp. $X_i(0)$ and $X_i(1)$) are convex.*

Proof. a) $X(0) = \mathbb{R}^n$.
b) Suppose that $X(1) \neq \emptyset$. For $x_i \in X(1)$, $i = 1, 2$, define

$$\Omega_i = \{\omega \mid A(\omega) x_i \ge b(\omega)\}, \qquad i = 1, 2.$$

Then

$$P_\omega(\Omega_i) = 1, \qquad i = 1, 2, \quad \text{and}$$

$$P_\omega(\Omega_1 \cap \Omega_2) = 1 \quad \text{(see proof of Th. III.4)},$$

and

$$\text{for} \quad \omega \in \Omega_1 \cap \Omega_2$$

$$A(\omega) x_i \ge b(\omega)$$

and therefore

$$A(\omega)\big(\lambda x_1 + (1-\lambda) x_2\big) \ge b(\omega) \quad \text{for} \quad 0 \le \lambda \le 1.$$

Hence for $\lambda \in [0,1]$

$$\Omega_\lambda = \{\omega \mid A(\omega)\big(\lambda x_1 + (1-\lambda) x_2\big) \ge b(\omega)\} \supset \Omega_1 \cap \Omega_2,$$

implying

$$\lambda x_1 + (1-\lambda) x_2 \in X(1) \quad \text{for} \quad \lambda \in [0,1]. \ \square$$

Th. 1 may obviously be restated in the following way:

Corollary 2. Given P_ω there are real numbers α^0 and α_i^0 such that $X(\alpha)$ and $X_i(\alpha_i)$ are convex sets for $\alpha \geq \alpha^0$ and $\alpha_i \geq \alpha_i^0$.

According to Example 1 and Cor. 2 one has to determine α^0 resp. α_i^0 for each particular probability distribution P_ω. Among others who were concerned with these problems, Marti [10] found the results Th. 3 and Th. 5.

Theorem 3. *Let P_ω be a finite discrete probability distribution, i.e.*

$$p_i = P_\omega(\omega_i) > 0, \quad i = 1, \ldots, r \quad \textit{and} \quad \sum_{i=1}^{r} p_i = 1.$$

For every $\alpha^0 > \max\limits_{1 \leq i \leq r} (1 - p_i)$ *and* $\alpha_i^0 > \max\limits_{1 \leq i \leq r} (1 - p_i)$

the sets $X(\alpha)$ and $X_i(\alpha_i)$ are convex for $\alpha > \alpha^0$ resp. $\alpha_i > \alpha_i^0$.

Proof. For $N = \{1, \ldots, r\}$ and $I \subset N$, $I \neq N$

$$\sum_{i \in I} p_i \leq 1 - p_j \quad \forall j \in N - I$$
$$\leq \max_{j \in N - I} (1 - p_j) \leq \max_{j \in N} (1 - p_j).$$

Hence for $I \subset N$

$$\sum_{i \in I} p_i > \max_{j \in N} (1 - p_j)$$

implies

$$\sum_{i \in I} p_i = 1.$$

This yields for $x \in X(\alpha)$, where $\alpha > \max\limits_{1 \leq i \leq r} (1 - p_i)$,

$$P_\omega(\{\omega \mid A(\omega) x \geq b(\omega)\}) = \sum_{A(\omega_i) x \geq b(\omega_i)} p_i \geq \alpha$$

and hence

$$P_\omega(\{\omega \mid A(\omega) x \geq b(\omega)\}) = 1;$$

and this implies $X(\alpha) = X(1)$, which is convex by Th. 1. □

For finite discrete distributions, the condition $\alpha > \max\limits_{1 \leq i \leq r} (1 - p_i)$ is, by Th. 3, sufficient for convexity but not necessary, as may be seen in Example 1, where $X(\alpha)$ is convex for $\alpha > \frac{1}{3}$, but $\max\limits_{i} (1 - p_i) = \frac{2}{3}$. However, the condition cannot be weakened in general, as the following example demonstrates.

Example 2. Let P_ω be a discrete distribution so that

$$p_1 = P_\omega(\omega_1) = \frac{1}{4}; \quad p_2 = P_\omega(\omega_2) = \frac{1}{2}; \quad p_3 = P_\omega(\omega_3) = \frac{1}{4}.$$

Let

$$A(\omega_1) = \begin{pmatrix} 1 & -1 \\ 0 & 1 \end{pmatrix}, \quad A(\omega_2) = \begin{pmatrix} 1 & -1 \\ -2 & -3 \end{pmatrix}, \quad A(\omega_3) = \begin{pmatrix} -1 & -1 \\ -1 & 3 \end{pmatrix}$$

and

$$b(\omega_1) = \begin{pmatrix} -2 \\ 3 \end{pmatrix}, \quad b(\omega_2) = \begin{pmatrix} 0 \\ -25 \end{pmatrix}, \quad b(\omega_3) = \begin{pmatrix} -8 \\ 0 \end{pmatrix}.$$

If

$$K(\omega_1)=\{(\xi,\eta)\in\mathbb{R}^2|\xi-\eta\geq -2;\quad \eta\geq 3\}$$
$$K(\omega_2)=\{(\xi,\eta)\in\mathbb{R}^2|\xi-\eta\geq 0;\quad -2\xi-3\eta\geq -25\}$$
$$K(\omega_3)=\{(\xi,\eta)\in\mathbb{R}^2|-\xi-\eta\geq -8;\quad -\xi+3\eta\geq 0\},$$

then

$$P_\omega(\{\omega|A(\omega)x\geq b(\omega)\})=\sum_{i\in I(x)} p_i,$$

where

$$I(x)=\{i|x\in K(\omega_i)\}.$$

Since

$$\max_i(1-p_i)=\frac{3}{4},$$

we know that, for

$$\alpha>\frac{3}{4},\quad X(\alpha)=X(1)\quad \text{is convex.}$$

Here $X(1)=K(\omega_1)\cap K(\omega_2)\cap K(\omega_3)$ is the triangle with the vertices (3, 3), (5, 3), (4, 4).

But for $\alpha=\frac{3}{4}$ we get

$$X\left(\frac{3}{4}\right)=[K(\omega_1)\cap K(\omega_2)]\cup[K(\omega_2)\cap K(\omega_3)]$$

which is not convex, because

$$x=(6,2)\in K(\omega_2)\cap K(\omega_3)$$
$$y=(6,4)\in K(\omega_1)\cap K(\omega_2)$$

and therefore

$$x\in X\left(\frac{3}{4}\right)\quad\text{and}\quad y\in X\left(\frac{3}{4}\right).$$

But for

$$z=\frac{3}{4}x+\frac{1}{4}y=\left(6,\frac{5}{2}\right)\quad\text{we have}$$

$$z\in K(\omega_2),\quad z\notin K(\omega_1),\quad z\notin K(\omega_3)$$

and hence

$$z\notin X\left(\frac{3}{4}\right).$$

In this example we have made use of the fact that $\max_{1\leq i\leq r}(1-p_i)$ is not unique. If we have a discrete distribution such that $\min_i p_i$ is uniquely determined, we may decrease the lower bound of the probability level given in Th. 3.

Theorem 4. *Let P_ω be a finite discrete probability distribution, i.e.*

$p_i = P_\omega(\omega_i) > 0$, $i \in N = \{1,\ldots,r\}$, *and* $\sum_{i=1}^{r} p_i = 1$, so *that* $\min_{i \in N} p_i = p_{i_0}$ *is uniquely determined. Then the sets* $X(\alpha)$ *and* $X_i(\alpha)$ *are convex for every* $\alpha > 1 - p_{i_1}$, *where*

$$p_{i_1} = \min_{i \in N - \{i_0\}} p_i.$$

Proof. For $I \subset N$ we have

$$\sum_{i \in I} p_i \begin{cases} = 1 & \text{if} \quad I = N \\ \leq 1 - p_{i_0} & \text{if} \quad i_0 \notin I \\ \leq 1 - p_{i_1} & \text{if} \quad j \notin I \quad j \neq i_0. \end{cases}$$

Hence

$$\sum_{i \in I} p_i > 1 - p_{i_1} \quad \text{implies} \quad I \supset N - \{i_0\}.$$

With $K(\omega_i) = \{x \mid A(\omega_i)x \geq b(\omega_i)\}$ it follows immediately that

$$X(\alpha) = \bigcap_{i \in N - \{i_0\}} K(\omega_i) \quad \text{for} \quad 1 - p_{i_1} < \alpha \leq 1 - p_{i_0} \quad \text{and}$$

$$X(\alpha) = \bigcap_{i \in N} K(\omega_i) \quad \text{for} \quad \alpha > 1 - p_{i_0},$$

which yields the theorem, since every $K(\omega_i)$ is a convex polyhedral set. □

The situation described in this theorem can be observed in Example 1, where

$$p_{i_0} = p_1 = \frac{1}{3}, \quad p_{i_1} = p_2 = \frac{2}{3}$$

and where $X(\alpha)$ is in fact convex for $\alpha > (1 - p_{i_1}) = \frac{1}{3}$.

Besides these convexity statements on $X(\alpha)$ in the discrete distribution case, the convexity of $X_i(\alpha)$ only seems to be investigated for some special distributions as long as $A(\omega)$ is random.

Theorem 5. *Suppose that the random variables* $a_{i1}(\omega), a_{i2}(\omega), \ldots, a_{in}(\omega), b_i(\omega)$ *have a joint* $(n+1)$*-dimensional normal distribution. Then* $X_i(\alpha_i)$ *is convex for* $\alpha_i \geq \frac{1}{2}$.

Proof. If d and f are $(n+1)$-dimensional random vectors with probability density functions $\varphi(\xi)$ and $\psi(\zeta)$ respectively, and if $f = Td$, where T is a nonsingular $(n+1) \times (n+1)$ matrix, then

$$\psi(\zeta) = \varphi(T^{-1}\zeta) \cdot |\det T|^{-1}.$$

d has a normal distribution if

$$\varphi(\xi) = \gamma e^{-\frac{1}{2}(\xi - m)' S (\xi - m)}$$

where γ is a constant such that

$$\int_{\mathbb{R}^{n+1}} \varphi(\xi) \mathrm{d}\xi = 1,$$

S and hence S^{-1} are symmetric and strictly positive definite, and

$$\mu_i = m_i = E(d_i)$$
$$\sigma_i^2 = (S^{-1})_{ii} = E(d_i - \mu_i)^2.$$

Suppose now, that

$$d' = (a_{i1}, a_{i2}, \ldots, a_{in}, b_i),$$

and for some $x \in \mathbb{R}^n$

$$T(x) = \begin{pmatrix} 1 & 0 & \ldots\ldots\ldots\ldots\ldots\ldots & 0 \\ 0 & 1 & 0 \ldots\ldots\ldots\ldots\ldots & 0 \\ \vdots & & & \\ -x_1 & -x_2 \ldots\ldots\ldots & -x_n & 1 \end{pmatrix}.$$

If $f = T(x)d$ then

$$f_i = d_i, \quad 1 \le i \le n,$$
$$f_{n+1} = d_{n+1} - \sum_{i=1}^{n} x_i d_i = b_i - \sum_{j=1}^{n} a_{ij} x_j,$$

and f has the density

$$\begin{aligned} \psi(\zeta) &= \varphi(T^{-1}(x)\zeta), \quad \text{since } \det T(x) = 1 \\ &= \gamma e^{\frac{1}{2}(T^{-1}(x)\zeta - m)' S (T^{-1}(x)\zeta - m)} \\ &= \gamma e^{\frac{1}{2}(\zeta - r)' Q(x)(\zeta - r)} \end{aligned}$$

where

$$r = T(x) m$$
$$Q(x) = T^{-1}(x)' S T^{-1}(x)$$
$$Q^{-1}(x) = T(x) S^{-1} T(x)'.$$

Hence f has a normal distribution and especially

$$f_{n+1} = b_i - \sum_{j=1}^{n} a_{ij} x_j$$

has mean value

$$\mu_{n+1}(x) = r_{n+1} = m_{n+1} - \sum_{i=1}^{n} m_i x_i$$

and variance

$$\hat{\sigma}_{n+1}^2(x) = (Q^{-1}(x))_{n+1,n+1} = (-x_1 - x_2 \ldots -x_n 1) S^{-1} \begin{pmatrix} -x_1 \\ -x_2 \\ \vdots \\ -x_n \\ 1 \end{pmatrix}.$$

Since S^{-1} is positive definite, it is easily shown that $\hat{\sigma}_{n+1}(x)$ is convex in x ($\hat{\sigma}_{n+1}(x)$ may be regarded as a norm of the vector $(-x_1 - x_2 \ldots -x_n 1)'$).

Obviously

$$\frac{f_{n+1} - \hat{\mu}_{n+1}(x)}{\hat{\sigma}_{n+1}(x)}$$

has the standard normal distribution with mean value 0 and variance 1, whose distribution function shall be called $\Phi(\tau)$. Now it is evident that

$$P\left(\sum_{j=1}^{n} a_{ij}x_j \geq b_i\right) = P(f_{n+1} \leq 0)$$
$$= P\left(\frac{f_{n+1} - \hat{\mu}_{n+1}(x)}{\hat{\sigma}_{n+1}(x)} \leq -\frac{\hat{\mu}_{n+1}(x)}{\hat{\sigma}_{n+1}(x)}\right) = \Phi\left(-\frac{\hat{\mu}_{n+1}(x)}{\hat{\sigma}_{n+1}(x)}\right)$$

and

$$\Phi\left(\frac{-\hat{\mu}_{n+1}(x)}{\hat{\sigma}_{n+1}(x)}\right) \geq \alpha_i$$

if and only if

$$\Phi^{-1}(\alpha_i) \cdot \hat{\sigma}_{n+1}(x) + \hat{\mu}_{n+1}(x) \leq 0.$$

Since $\hat{\mu}_{n+1}(x)$ and $\hat{\sigma}_{n+1}(x)$ are convex in x, this inequality describes a convex set as long as $\Phi^{-1}(\alpha_i) \geq 0$, which is true for $\alpha_i \geq \frac{1}{2}$. □

The following example shows that the convexity of $X_i(\alpha_i)$ is in general not maintained for $\alpha_i < \frac{1}{2}$.

Example 3. Suppose that a and b are independent random variables with normal distributions such that

$$m_1 = E(a) = 1; \quad \sigma_1^2 = E(a - m_1)^2 = 3$$
$$m_2 = E(b) = 2; \quad \sigma_2^2 = E(b - m_2)^2 = 1.$$

Then

$$P(b - xa \leq 0) = \Phi\left(-(2-x)(1+3x^2)^{-\frac{1}{2}}\right) = \begin{cases} \Phi(-6/7) & \text{for} \quad x = -4 \\ \Phi(-2) & \text{for} \quad x = 0 \\ \Phi(2/7) & \text{for} \quad x = 4. \end{cases}$$

For $\alpha = \Phi\left(\frac{-6}{7}\right) < \frac{1}{2}$ we get

$P(b - xa \leq 0) \geq \alpha$ for $x = +4$ and $x = -4$, but not for $x = 0$.

As a matter of fact, for $x \in \mathbb{R}$ the results $X_i(\alpha_i) = \mathbb{R}$ and $X_i(\alpha_i) \neq \mathbb{R}$ with $X_i(\alpha_i)$ convex are also possible:

Example 4. Assume, as in Example 3, that a and b are independently normally distributed such that $m_1 = m_2 = 1$;

$$\sigma_1^2 = \sigma_2^2 = 1 \quad \text{and} \quad \alpha = \Phi(-1) < \frac{1}{2}.$$

Then

$$\Phi^{-1}(\alpha)\hat{\sigma}_2(x)+\hat{\mu}_2(x)=-\sqrt{x^2+1}+1-x\begin{cases}>0 & \text{for} \quad x<0\\ \leq 0 & \text{for} \quad x\geq 0\end{cases}$$

and hence $X(\alpha)=\{x \mid x\geq 0\}\neq \mathbb{R}$.

Example 5. If under the same assumption as in Example 4

$$m_1=m_2=1; \quad \sigma_1^2=2; \quad \sigma_2^2=4 \quad \text{and} \quad \alpha=\Phi(-1),$$

then

$$\Phi^{-1}(\alpha)\hat{\sigma}_2(x)+\hat{\mu}_2(x)=-\sqrt{4+2x^2}+1-x<0 \quad \text{for all} \quad x\in\mathbb{R}$$

and hence $X(\alpha)=\mathbb{R}$.

It turns out that a result as in Example 4 is only possible in $\mathbb{R}$.

Theorem 6. *Suppose that the random variables* $a_{i1}(\omega), a_{i2}(\omega), \ldots, a_{in}(\omega), b_i(\omega)$, *where* $n>1$, *have a joint* $(n+1)$*-dimensional normal distribution. If* $0<\alpha_i<\frac{1}{2}$, *then either* $X_i(\alpha_i)=\mathbb{R}^n$ *or* $X_i(\alpha_i)$ *is a nonempty nonconvex set.*

Proof. From Th. 5 we know that $x\in X_i(\alpha_i)$ if and only if

$$\Phi^{-1}(\alpha_i)\hat{\sigma}_{n+1}(x)+\hat{\mu}_{n+1}(x)\leq 0$$

where

$$\sigma_{n+1}^2(x)=\tilde{x}'S^{-1}\tilde{x}, \quad \text{if} \quad \tilde{x}=\begin{pmatrix}-x\\ 1\end{pmatrix}\in\mathbb{R}^{n+1}$$

and $\hat{\mu}_{n+1}(x)=m_{n+1}-m'x$, if $m'=(m_1,\ldots,m_n)$. Suppose that $X_i(\alpha_i)\neq\mathbb{R}^n$ and $x\notin X_i(\alpha_i)$. Since $n>1$ there exists a $y\in\mathbb{R}^n$ such that $y\neq 0$ and $m'y=0$. If we

define $\tilde{y}=\begin{pmatrix}-y\\ 0\end{pmatrix}$, then, for $\lambda\in\mathbb{R}$,

$$\hat{\mu}_{n+1}(x+\lambda y)=\hat{\mu}_{n+1}(x)$$

and

$$\hat{\sigma}_{n+1}^2(x+\lambda y)=\tilde{x}'S^{-1}\tilde{x}+2\lambda\tilde{x}'S^{-1}\tilde{y}+\lambda^2\tilde{y}'S^{-1}\tilde{y}.$$

Since $\tilde{y}\neq 0, \Phi^{-1}(\alpha_i)<0$ and S^{-1} is positive definite, there exists a $\lambda_0\in\mathbb{R}$, $\lambda_0\neq 0$ such that

$$[\Phi^{-1}(\alpha_i)]^2\hat{\sigma}_{n+1}^2(x+\lambda_0 y)>\hat{\mu}_{n+1}^2(x)=\hat{\mu}_{n+1}^2(x+\lambda_0 y)$$

and

$$[\Phi^{-1}(\alpha_i)]^2\hat{\sigma}_{n+1}^2(x-\lambda_0 y)>\hat{\mu}_{n+1}^2(x)=\hat{\mu}_{n+1}^2(x-\lambda_0 y).$$

Hence

$$x+\lambda_0 y\in X_i(\alpha_i) \quad \text{and} \quad x-\lambda_0 y\in X_i(\alpha_i),$$

whereas

$$x=\frac{1}{2}(x+\lambda_0 y)+\frac{1}{2}(x-\lambda_0 y)\notin X_i(\alpha_i),$$

and therefore $X_i(\alpha_i)$ is nonempty and nonconvex. □

A result similar to Th. 5 was obtained by K. Marti [10] for a joint Cauchy distribution.
Except in the case of a finite discrete distribution we have presented convexity statements only for $X_i(\alpha_i)$, but not for $X(\alpha)$. This corresponds to the state of research in the field, as long as the matrix A is supposed to be random. However, for fixed matrices A — i.e. only the right hand side $b(\omega)$ is random — $X_i(\alpha_i)$ is always convex, and the convexity of $X(\alpha)$ has recently been proved by A. Prékopa [14] for a rather broad class of probability distributions, including the normal distribution.

Theorem 7. *Suppose that A is fixed and $b(\omega)$ is random.*
Then

$$X_i(\alpha_i)=\{x \mid P_\omega(\{\omega \mid A_i x \geq b_i(\omega)\})\geq \alpha_i\}$$

is convex for every probability distribution of $b(\omega)$.

Proof. Let $F_i(\tau)$ be the distribution function of $b_i(\omega)$.
Then $x^j \in X_i(\alpha_i)$, $j=1,2$, if and only if $F_i(A_i x^j)\geq \alpha_i$, $j=1,2$. For $\hat{x}=\lambda x^1+(1-\lambda)x^2, \lambda\in(0,1)$, we get

$$\begin{aligned} F_i(A_i\hat{x}) &\geq F_i\big(\min_{j=1,2} A_i x^j\big) \\ &= \min_{j=1,2}\{F_i(A_i x^j)\}\geq \alpha_i, \end{aligned}$$

by the monotonicity of distribution functions.
Hence $X_i(\alpha_i)$ is convex. □

It is much more difficult to get convexity statements for $X(\alpha)$. If $F(z)$ is the distribution function of $b(\omega)$, then

$$X(\alpha)=\{x \mid F(Ax)\geq \alpha\}.$$

Therefore, if A has full rank, $X(\alpha)$ is convex for every $\alpha\in[0,1]$ if and only if $F(z)$ is quasi concave, i.e.

$$F\left(\lambda z^1+(1-\lambda)z^2\right)\geq \min\{F(z^1),F(z^2)\} \quad \text{for all} \quad z^j\in\mathbb{R}^m \quad \text{and} \quad \lambda\in(0,1).$$

Although every distribution function of a one dimensional random variable is quasi concave because of the monotonicity of distribution functions, this is in general not true for multivariate distribution functions as the following example shows:
Example 6. Let $b(\omega)$ be a two-dimensional random variable with the discrete distribution

$$P\begin{pmatrix}2\\0\end{pmatrix}=P\begin{pmatrix}0\\2\end{pmatrix}=\frac{1}{2}.$$

Then for $F(z)=P(b\leq z)$ we have

$$F\begin{pmatrix}2\\0\end{pmatrix}=F\begin{pmatrix}0\\2\end{pmatrix}=\frac{1}{2}, \quad \text{but} \quad F\begin{pmatrix}1\\1\end{pmatrix}=0,$$

which shows that $F(z)$ is not quasi concave.

If the probability measure P defined on $\mathbb{R}^n$ has a density function $f(x)$ (with respect to the Lebesgue measure μ on $\mathbb{R}^n$), the problem arises under which conditions on $f(x)$ the measure P is quasi concave. P is quasi concave if, for

$$\lambda\mathfrak{A}+(1-\lambda)\mathfrak{B}=\{z\,|\,z=\lambda x+(1-\lambda)y, x\in\mathfrak{A}, y\in\mathfrak{B}\},$$
$$P(\lambda\mathfrak{A}+(1-\lambda)\mathfrak{B})\geq\min\{P(\mathfrak{A}),P(\mathfrak{B})\}$$

for all convex subsets $\mathfrak{A}$ and $\mathfrak{B}$ of $\mathbb{R}^n$ and all $\lambda\in(0,1)$.)
Obviously the distribution function of a quasi concave probability measure is quasi concave.

We call a density function $f(x)$ almost quasi concave, if for every $a\in\mathbb{R}^n$ and $b\in\mathbb{R}^n$ such that $a=-\gamma b$ and $\gamma>0$, $f(x)\geq\min\{f(x+a),f(x+b)\}$ almost everywhere with respect to μ. Then we may state

Theorem 8. *Let P be a quasi concave probability measure on $\mathbb{R}^n$ with the continuous density function $f(x)$. Then $f(x)$ is almost quasi concave.*

Proof. Suppose $f(x)$ were not almost quasi concave. Then there exist
$a\in\mathbb{R}^n$, $b\in\mathbb{R}^n$, $\gamma>0$ with $a=-\gamma b$ such that $\mu(\{x|f(x)<\min[f(x+a),f(x+b)]\})>0$.
Thus, there exists a convex Borel measurable set $\mathfrak{K}$ (for example a sphere) such that

$$\mathfrak{K}\subset\{x|f(x)<\min[f(x+a),f(x+b)]\}\quad\text{and}\quad\mu(\mathfrak{K})>0.$$

Then

$$\begin{aligned}P(\mathfrak{K})=\int_{\mathfrak{K}}f(z)\mathrm{d}\mu(z)&<\int_{\mathfrak{K}}\min[f(z+a),f(z+b)]\mathrm{d}\mu(z)\\&\leq\min\Big[\int_{\mathfrak{K}}f(z+a)\mathrm{d}z;\int_{\mathfrak{K}}f(z+b)\mathrm{d}z\Big]\\&=\min[P(\mathfrak{K}+a);P(\mathfrak{K}+b)]\end{aligned}$$

in contradiction to the quasi concavity of P, since

$$\mathfrak{K}=\lambda(\mathfrak{K}+a)+(1-\lambda)(\mathfrak{K}+b)\quad\text{with}\quad\lambda=\frac{1}{1+\gamma}\in(0,1).\ \square$$

However, almost quasi concavity of a density function does not in general imply quasi concavity of the related probability measure.

Example 7. Let $f(x)=\frac{1}{124}\varphi(x)$ be a density function on $\mathbb{R}^2$, where

$$\varphi(x)=\begin{cases}21 & \text{if}\quad x\in\mathfrak{M}=\{x|0\leq x_1\leq 1,\ -2\leq x_2\leq 0\}\\ 1 & \text{if}\quad x\in\mathfrak{N}=\{x|-11\leq x_1\leq 1,\ -6\leq x_2\leq 1, x\notin\mathfrak{M}\}\\ 0 & \text{else.}\end{cases}$$

Obviously $f(x)$ is quasi concave (and hence almost quasi concave). Take

$$z^1=(-1,1),\quad z^2=(1,-1)\quad\text{and}\quad\hat{z}=\frac{1}{2}z^1+\frac{1}{2}z^2=(0,0).$$

Then we have for the distribution function

$$F(z^1)=\int_{-\infty}^{z^1}f(x)\mathrm{d}x=\frac{70}{124}$$

$$F(z^2) = \int_{-\infty}^{z^2} f(x)\mathrm{d}x = \frac{80}{124}$$

$$F(\hat{z}) = \int_{-\infty}^{\hat{z}} f(x)\mathrm{d}x = \frac{66}{124} < F(z^1) < F(z^2).$$

Hence the distribution function and consequently the probability measure is not quasi concave.

With respect to sufficient conditions the strongest results known so far are due to A. Prékopa [14]. He was concerned with logarithmic concave measures, which due to their definition satisfy the inequality

$$P(\lambda\mathfrak{A} + (1-\lambda)\mathfrak{B}) \geq P^{\lambda}(\mathfrak{A}) \cdot P^{1-\lambda}(\mathfrak{B})$$

for all convex subsets $\mathfrak{A}$ and $\mathfrak{B}$ of $\mathbb{R}^n$ and all $\lambda \in (0,1)$.

Obviously a logarithmic concave probability measure is also quasi concave. The main result is based on

Theorem 9 (Prékopa's inequality). *Let f and g be nonnegative Borel measurable functions defined on $\mathbb{R}^n$ and let*

$$r(t) = \sup_{\lambda x + (1-\lambda)y = t} f(x) \cdot g(y); \quad t \in \mathbb{R}^n,$$

where λ is a constant, $0 < \lambda < 1$.
Then $r(t)$ is Borel measurable and the following inequality holds:

$$\int_{\mathbb{R}^n} r(t)\mathrm{d}t \geq \left(\int_{\mathbb{R}^n} f^{\frac{1}{\lambda}}(x)\mathrm{d}x \right)^{\lambda} \cdot \left(\int_{\mathbb{R}^n} g^{\frac{1}{1-\lambda}}(y)\mathrm{d}y \right)^{1-\lambda}.$$

The proof of this theorem can be found in A. Prékopa [13], [14] and L. Leindler [8].

Theorem 10. *Let $f(x)$ be a probability density function defined on $\mathbb{R}^n$, with the representation $f(x) = \gamma \cdot e^{-Q(x)}$, where $Q(x)$ is a convex function. Then the corresponding probability measure P is logarithmic concave.*

Proof. Let $\mathfrak{A}$ and $\mathfrak{B}$ be arbitrary convex subsets of $\mathbb{R}^n$ and let $\lambda \in (0,1)$. Define $f_i(x)$, $i = 1, 2, 3$, as follows:

$$f_1(x) = \begin{cases} f(x) & \text{if } x \in \mathfrak{A} \\ 0 & \text{otherwise} \end{cases}$$

$$f_2(x) = \begin{cases} f(x) & \text{if } x \in \mathfrak{B} \\ 0 & \text{otherwise} \end{cases}$$

$$f_3(x) = \begin{cases} f(x) & \text{if } x \in \lambda\mathfrak{A} + (1-\lambda)\mathfrak{B} \\ 0 & \text{otherwise.} \end{cases}$$

For every $x \in \lambda\mathfrak{A} + (1-\lambda)\mathfrak{B}$ and every $y \in \mathfrak{A}$ and $z \in \mathfrak{B}$ such that $\lambda y + (1-\lambda)z = x$, in view of the convexity of $Q(x)$, we have

$$\begin{aligned} f(x) = \gamma e^{-Q(x)} &\geq \gamma e^{-\lambda Q(y) - (1-\lambda)Q(z)} \\ &= f^{\lambda}(y) \cdot f^{(1-\lambda)}(z), \end{aligned}$$

implying immediately

$$f_3(x) \geq \sup_{\lambda y+(1-\lambda)z=x} f_1^{\lambda}(y) f_2^{(1-\lambda)}(z).$$

Now Th. 9 yields

$$\begin{aligned} P(\lambda\mathfrak{A}+(1-\lambda)\mathfrak{B}) &= \int_{\mathbb{R}^n} f_3(x)\mathrm{d}x \geq \int_{\mathbb{R}^n} \{\sup_{\lambda y+(1-\lambda)z=x} f_1^{\lambda}(y) f_2^{(1-\lambda)}(z)\}\mathrm{d}x \\ &\geq \left(\int_{\mathbb{R}^n} f_1(y)\mathrm{d}y\right)^{\lambda}\left(\int_{\mathbb{R}^n} f_2(z)\mathrm{d}z\right)^{(1-\lambda)} \\ &= P^{\lambda}(\mathfrak{A})\cdot P^{1-\lambda}(\mathfrak{B}), \end{aligned}$$

which establishes the logarithmic concavity of P. □

The convexity statements and examples given in this section show quite clearly that *in general* it is not yet known under which conditions a chance constrained program is a convex program. Moreover — even if convexity can be asserted — there are still considerable computational difficulties.

2. Relationship between Chance Constrained Programs and Two-Stage-Problems

We cannot in general expect that chance constrained programs and two-stage problems replace each other, because in practical situations it sometimes seems appropriate to require a probability level of feasibility but impossible to specify penalty costs for infeasibility and vice versa. And from the theory developed so far, we may suspect that chance constrained programs and two-stage problems are in general not equivalent, since, in general, chance constrained programs may be nonconvex whereas two-stage problems are always convex. Nevertheless there are relations between these two types of problems, which may at least help us to get more insight.

On the one hand under the assumptions of Th. III. 12 any fixed recourse problem with deterministic penalty costs q is equivalent to finding feasible points of generalized chance constraints. Here generalized chance constraints are constraints involving functions of the type $g(x)=P(Ax\geq b)$. If we start with the two-stage problem

$$\begin{aligned} &\min c'x+Q(x) \\ &\text{subject to } Tx=d \\ &\qquad x\geq 0 \end{aligned}$$

where T, d are deterministic and $Q(x)=E_{P_\omega}Q(x,\omega)$, and

$$\begin{aligned} &Q(x,\omega)=\min q'y \\ &\text{subject to } Wy=b(\omega)-A(\omega)x \\ &\qquad y\geq 0, \end{aligned}$$

then, according to the Kuhn-Tucker-theorem, we have to find a feasible solution of

$$\begin{aligned} c+\nabla Q(x)-T'u &\geq 0 \\ x'(c+\nabla Q(x)-T'u) &= 0 \\ Tx &= d \\ x &\geq 0. \end{aligned}$$

From Th. III.12 we know that

$$\nabla Q(x) = -\sum_{i=1}^{r} \int_{\mathfrak{B}_i(x)} (\tilde{q}_i' B_i^{-1} A(\omega))' \mathrm{d}P_\omega$$

where B_i, $i=1,\ldots,r$, are optimal bases of W (i.e. fulfil the simplex optimality condition) and

$$\mathfrak{B}_i(x) = \{\omega \mid B_i^{-1}(b(\omega)-A(\omega)x) > 0\} - \bigcup_{j=1}^{i-1} \mathfrak{B}_j(x).$$

Rewriting the integrals in $\nabla Q(x)$, using the conditional probability of b given A, yields generalized chance constraints. This fact becomes still more evident if $A(\omega) \equiv A$ is deterministic and the optimal feasible basis of the recourse program is determined uniquely almost everywhere (as for example in the simple recourse case). Then

$$\nabla Q(x) = -\sum_{i=1}^{r} (\tilde{q}_i' B_i^{-1} A)' P(B_i^{-1}(b(\omega)-Ax) > 0).$$

On the other hand we may sometimes use simple recourse problems to find feasible solutions of chance constraints. Consider the special simple recourse problem ($q_i^+ = \varrho > 0$, $q_i^- = 0$)

$$\psi(\varrho) = \operatorname*{Min}_{x\in X} \{c'x + \sum_{i=1}^{m} \varrho \int_{(b(\omega)-A(\omega)x)_i > 0} (b(\omega)-A(\omega)x)_i \mathrm{d}P_\omega\}$$

and let, for every $\varrho > 0$, $x(\varrho)$ be a solution.

Theorem 11. *Let X be compact. Then*

$$\mathfrak{B} = X \cap \{x \mid P_\omega(A(\omega)x \geq b(\omega)) = 1\} \neq \emptyset$$

if and only if

$$\lim_{\varrho\to\infty} \psi(\varrho) < \infty.$$

Proof. a) The condition is necessary, since obviously $\psi(\varrho)$ is monotone increasing, and if $\bar{x} \in \mathfrak{B}$, $\psi(\varrho) \leq c'\bar{x}$, because for $\bar{x} \in \mathfrak{B}$ $P_\omega((b(\omega)-A(\omega)\bar{x})_i > 0) = 0$ for $i=1,\ldots,m$.

b) The condition is also sufficient. Consider the functions

$$\varphi_{i\varrho}(\omega) = \begin{cases} 0 & \text{if } (b(\omega)-A(\omega)x(\varrho))_i \leq 0 \\ (b(\omega)-A(\omega)x(\varrho))_i & \text{otherwise.} \end{cases}$$

We have to show that

$$\lim_{\varrho\to\infty} P_\omega(\{\omega \mid \varphi_{i\varrho}(\omega) \geq \varepsilon\}) = 0 \quad \text{for all} \quad \varepsilon > 0 \quad \text{and} \quad i=1,\ldots,m.$$

i.e. that the functions $\varphi_{i\varrho}$, $i=1,\ldots,m$ converge in measure to zero.

Suppose, on the other hand, that for some i there exist $\varepsilon > 0$ and $\delta > 0$ such that for every $\varrho > 0$ there is $\Delta > 0$ for which

$$P_\omega(\{\omega \mid \varphi_{i\varrho+\Delta}(\omega) \geq \varepsilon\}) \geq \delta.$$

Then

$$\psi(\varrho+\Delta) \geq \alpha + \varepsilon \cdot \delta \cdot (\varrho + \Delta)$$

where $\alpha = \min_{x \in X} c'x$. This inequality contradicts the assumption that $\lim_{\varrho \to \infty} \psi(\varrho) < \infty$. Now, if $\{\varrho\}$ is some sequence increasing to ∞, there is a subsequence $\{\varrho_\nu\}$ such that $\varphi_{i\varrho_\nu}$, $i=1,\ldots,m$, converge to zero almost surely, (i.e. almost everywhere with respect to P_ω) and $\lim_{\nu \to \infty} x(\varrho_\nu) = x^* \in X$.

Therefore

$$P_\omega(\{\omega \mid \lim_{\nu \to \infty} \varphi_{i\varrho_\nu}(\omega) > 0\}) = P_\omega(\{\omega \mid (b(\omega) - A(\omega)x^*)_i > 0\}) = 0, \qquad i=1,\ldots,m,$$

yielding

$$P_\omega(\{\omega \mid A(\omega)x^* \ngeq b(\omega)\}) = P_\omega\left(\bigcup_{i=1}^{m} \{\omega \mid (b(\omega) - A(\omega)x^*)_i > 0\}\right)$$

$$\leq \sum_{i=1}^{m} P_\omega(\{\omega \mid (b(\omega) - A(\omega)x)_i > 0\}) = 0,$$

i.e. $P_\omega(\{\omega \mid A(\omega)x^* \geq b(\omega)\}) = 1.$ □

According to this theorem one may try to get feasible solutions of chance constraints by solving the parametric simple recourse problem mentioned above. However, one should be aware of the fact that the theorem could only be proved under the assumption that probability 1 could be attained.

References

[1] Avriel, M. and A.C. Williams: The Value of Information and Stochastic Programming. Operations Research **18**, 947–954 (1970).

[2] Beale, E.M.: The Use of Quadratic Programming in Stochastic Linear Programming. RAND P-2404, August 1961.

[3] Bereanu, B.: On Stochastic Linear Programming Distribution Problems, Stochastic Technology Matrix. Z. Wahrscheinlichkeitstheorie u verw. Geb. **8**, 148–152 (1967).

[4] Bereanu, B.: The Distribution Problem in Stochastic Linear Programming. The Cartesian Integration Method. Reprint No. 7103, Center of Mathematical Statistics of the Academy of the Socialist Republic of Romania, Bucharest (1971).

[5] Kall, P.: Qualitative Aussagen zu einigen Problemen der stochastischen Programmierung. Z. Wahrscheinlichkeitstheorie u. verw. Geb. **6**, 246–272 (1966).

[6] Kall, P.: Das zweistufige Problem der stochastischen linearen Programmierung. Z. Wahrscheinlichkeitstheorie u. verw. Geb. **8**, 101–112 (1967).

[7] Kall, P.: Some Remarks on the Distribution Problem of Stochastic Linear Programming. Methods of Operations Research, Meisenheim, **16**, 189–196 (1973).

[8] Leindler, L.: On a Certain Converse of Hölder's Inequality. Acta Scientiarum Mathematicarum, Szeged, **33**, 217–223 (1972).

[9] Madansky, A.: Inequalities for Stochastic Linear Programming Problems. Management Sci. **6**, 197–204 (1960).

[10] Marti, K.: Konvexitätsaussagen zum linearen stochastischen Optimierungsproblem. Z. Wahrscheinlichkeitstheorie u. verw. Geb. **18**, 159–166 (1971).

[11] Marti, K.: Entscheidungsprobleme mit linearem Aktionen- und Ergebnisraum. Z. Wahrscheinlichkeitstheorie u. verw. Geb. **23**, 133–147 (1972).

[12] Marti, K.: Über ein Verfahren zur Lösung einer Klasse linearer Entscheidungsprobleme. Z. Angew. Math. u. Mech., to appear.

[13] Prékopa, A.: Logarithmic Concave Measures with Applications to Stochastic Programming. Acta Scientiarum Mathematicarum, Szeged, **32**, 301–316 (1971).

[14] Prékopa, A.: On Logarithmic Concave Measures and Functions. Acta Scientiarum Mathematicarum, Szeged, **34**, 335–343 (1973).

[15] Walkup, D.W. and R.J.B. Wets: Stochastic Programs with Recourse. SIAM J. Appl. Math. **15**, 1299–1314 (1967).

[16] Wets, R.: Programming under Uncertainty: The Complete Problem. Z. Wahrscheinlichkeitstheorie **4**, 316–339 (1966).

[17] Wets, R.: Characterization Theorems for Stochastic Programs. Mathematical Programming **2**, 166–175 (1972).

[18] Wessels, J.: Stochastic Programming. Statistica Neerlandica **21**, 39–53 (1967).

[19] Kosmol, P.: Algorithmen zur konvexen Optimierung. OR-Verfahren, Band XVIII, 176–186 (1974).

[20] Kall, P.: Approximations to Stochastic Programs with Complete Fixed Recourse. Numer. Math. **22**, 333–339 (1974).

Books on Stochastic Programming

Faber, M.M.: Stochastisches Programmieren. Physica-Verlag, Würzburg, Wien, 1970.

Sengupta, J.K.: Stochastic Programming — Methods and Applications. North-Holland Publishing Company, Amsterdam, American Elsevier Publishing Company, Inc. New York, 1972.

Vajda, S.: Probabilistic Programming. Academic Press, New York and London, 1972.

Subject Index

Ökonometrie und Unternehmensforschung
Econometrics and Operations Research

Vol. I	H.P. Künzi, W. Krelle: Nichtlineare Programmierung Unter Mitwirkung von W. Oettli
Vol. II	G.B. Dantzig: Lineare Programmierung und Erweiterungen Ins Deutsche übertragen und bearbeitet von A. Jaeger
Vol. III	M. Girault: Stochastic Processes
Vol. IV	K.-H. Wolff: Methoden der Unternehmensforschung im Versicherungswesen
Vol. V	J.M. Danskin: The Theory of Max-Min and its Application to Weapons Allocation Problems
Vol. VI	H. Schneeweiß: Entscheidungskriterien bei Risiko
Vol. VII	P.L. Hammer (Ivanescu), S. Rudeanu: Boolean Methods in Operations Research and Related Areas With a preface by R. Bellman
Vol. VIII	Th. Marschak, Th.K. Glennan Jr., R. Summers: Strategy for R & D: Studies in the Microeconomics of Development
Vol. IX	M.J. Beckmann: Dynamic Programming of Economic Decisions
Vol. X	J. Schumann: Input-Output-Analyse
Vol. XI	W. Wittmann: Produktionstheorie
Vol. XII	W. Dinkelbach: Sensitivitätsanalysen und parametrische Programmierung
Vol. XIII	W. Knödel: Graphentheoretische Methoden und ihre Anwendungen
Vol. XIV	E. Nievergelt, O. Müller, F.E. Schlaepfer, W.H. Landis: Praktische Studien zur Unternehmensforschung
Vol. XV	H. Müller-Merbach: Optimale Reihenfolgen
Vol. XVI	R. Selten: Preispolitik der Mehrproduktenunternehmung in der statischen Theorie
Vol. XVII	L.P. Hyvärinen: Information Theory for Systems Engineers
Vol. XVIII	F.L. Wilke: Unternehmensforschung im Bergbau
Vol. XIX	J. Kornai: Anti-Äquilibrium. Über die Theorien der Wirtschaftssysteme und die damit verbundenen Forschungsaufgaben Übersetzer: L. Tarnóczy, M. Kenedi
Vol. XX	E. Blum, W. Oettli: Mathematische Optimierung. Grundlagen und Verfahren
Vol. XXI	P. Kall: Stochastic Linear Programming

A New Series

Applications of Mathematics

Subtitles: Applied Probability, Control, Economics, Information and Communication, Modeling and Identification, Numerical Techniques, Optimization

Editors: A. V. Balakrishnan, W. Hildenbrand

Applications of Mathematics will be devoted to publications in the new areas of applied mathematics i.e. those different from the natural sciences. In recent years the fields of applicability of modern mathematics have been expanded beyond the traditional boundaries; this has made new areas accessible like economics, biology, certain aspects of engineering, etc., and lead, concurrently, to the development of appropriate new methods and strategies (e.g. dynamic programming, stochastic differential equations, topological methods). *Applications of Mathematics* will be devoted to these developments; the publications in the series will cover fields where a mathematically meaningful, quantitative and qualitative approach has become possible, with emphasis on theoretically interesting, not too specialized presentations and mathematical thoroughness with specific, concrete questions as starting points.

The first four volumes:

Volume 1. W.H. Fleming and R.W. Rishel: Deterministic and Stochastic Optimal Control. 4 figures, XI, 222 pages. 1975. ISBN 3-540-90155-8.

Volume 2. G.I. Marchuk: Methods of Numerical Mathematics. VIII, 352 pages. ISBN 3-540-90156-6. Scheduled to appear in Dec. 1975/Jan. 1976

Volume 3. A.V. Balakrishnan: Applied Functional Analysis. Approx. 605 pages. ISBN 3-540-90157-4. Scheduled to appear in Jan. 1976

Volume 4. A.A. Borovkov: Stochastic Processes in Queueing Theory. 14 figures, approx. 250 pages. ISBN 0-387-90161-2. Scheduled to appear in Jan. 1976

Springer-Verlag Berlin Heidelberg New York